ATTITUDES TO ANIMALS:
VIEWS IN ANIMAL WELFARE

Attitudes to Animals provides a foundation that the reader can use to make ethical choices about animals. It will challenge readers to question their current views, attitudes and perspectives on animals and the nature of the human–animal relationship.

Human perspectives on the human–animal relationships reflect what we have learned, together with spoken and unspoken attitudes and assumptions, from our families, society, media, education and employment. This thought-provoking book will ask what it is to be human, what to be animal, and what is the nature of relationship between them. These questions are addressed through philosophical and ethical discussions, scientific evidence and dynamic theoretical approaches. *Attitudes to Animals* will also encourage us to think not only of our relationships to non-human animals, but also those with other, human, animals.

FRANCINE L. DOLINS has taught psychology and psychobiology at Centre College in Danville, Kentucky. In addition to a long-term interest in animal welfare issues, her research has focused on spatial cognition and foraging in non-human primates. She has worked with captive primate populations in the United Kingdom and the United States, and has also conducted field research in Peru, Costa Rica and Madagascar. Currently, Francine Dolins is the Program Officer for Research in the Animal Research Issues Section of The Humans Society of the United States.

For Christopher – for sharing his joy in all that is life

ATTITUDES TO ANIMALS: VIEWS IN ANIMAL WELFARE

Edited by

FRANCINE L. DOLINS

CAMBRIDGE
UNIVERSITY PRESS

PUBLISHED BY THE PRESS SYNDICATE OF THE UNIVERSITY OF CAMBRIDGE
The Pitt Building, Trumpington Street, Cambridge, United Kingdom

CAMBRIDGE UNIVERSITY PRESS
The Edinburgh Building, Cambridge CB2 2RU, UK http://wwww.cup.cam.ac.uk
40 West 20th Street, New York, NY 10011-4211, USA http://wwww.cup.org
10 Stamford Road, Oakleigh, Melbourne 3166, Australia

First published in 1999

Printed in the United Kingdom at the University Press, Cambridge

Typeset in Times New Roman 10/13pt [VN]

A catalogue record for this book is available from the British Library

Library of Congress Cataloguing in Publication Data
Dolins, Francine L. (Francine Leigh), 1964–
Attitudes to animals: views in animal welfare/Francine L. Dolins
p.   cm.
Includes bibliographical references and index.
ISBN 0 521 47342 X. – ISBN 0 521 47906 1 (pbk.)
1. Animal welfare – moral and ethical aspects. 2. Human–animal relationships – moral
and ethical aspects. I. Title
HV4708.D63   1999
179′.3—dc21   97-32155 CIP

ISBN 0 521 47342 X hardback
ISBN 0 521 47906 I paperback

Contents

Contributors

Charles Bergman
Department of English, Pacific Lutheran University, Tacoma, Washington, WA 98477, USA

Donald M. Broom
Department of Clinical Veterinary Medicine, University of Cambridge, Madingley Road, Cambridge CB3 0ES, UK

Richard W. Byrne
Scottish Primate Research Group, School of Psychology, University of St Andrews, St Andrews, Fife KY16 9JU, UK

David E. Cooper
Department of Philosophy, University of Durham, 50 Old Elvet Road, Durham DH1 3HN, UK

David Dewhurst
Faculty of Health and Environment, Leeds Metropolitan University, Calverley Street, Leeds LS1 3HE, UK

Francine L. Dolins
The Humane Society of the United States, Animal Issues Research Section, Animal Issues Research Section, 2100 L Street, NW, Washington, DC 20037, USA

Susan D. Healy
Department of Psychology, University of Newcastle-upon-Tyne, Newcastle-upon-Tyne NE1 7RU UK

M. Patricia Hindley
School of Communication, Simon Fraser University, Burnaby, British Columbia, V5A 1S6, Canada

Mary Midgley
1a Collingwood Terrace, Newcastle-upon-Tyne NE2 2JP, UK

James E. King
Department of Psychology, University of Arizona, Tucson, AZ 85721, USA

Mary Midgley
Phyllis Passariello
Anthropology Program, Centre College, Danville, 600 West Walnut Street, Danville, KY 40422, USA

Andrew J. Petto
Center for Science Education, PO Box 8880 National, Madison, WI 53708-8880, USA

Karla D. Russell
RRI, Box 87, Intervale, NH 03845, USA

James A. Serpell
Department of Clinical Studies, School of Veterinary Medicine, University of Pennsylvania, Philadelphia, PA 19104, USA

David Shepherdson
Metro Washington Park Zoo, Portland, Oregon, OR 97221, USA

Martin J. Tovée
Department of Psychology, University of Newcastle-upon-Tyne, Newcastle-upon-Tyne NE1 7RU, UK

Henk Verhoog
Theoretical Biology Division, Leiden University, Kaiserstraat 63, 2311 GP Leiden PO Box 9516, 2300 RA Leiden, The Netherlands

Françoise Wemelsfelder
Genetics and Behavioural Sciences Department, Scottish Agricultural College, Bush Estate, Penicuik EH26 0QE, UK

Robert J. Young
School of Agriculture and Horticulture, De Montfort University, Caythorpe, Grantham, Lincolnshire NG32 3EP, UK

Acknowledgements

There are many people both professionally and personally whom I would like to acknowledge and extend my deepest thanks to in helping to make this book possible. First, I would like to thank each of the contributors who gave their time and energy in researching and writing their chapters. I would like to thank our editor at Cambridge University Press, Dr Tracey Sanderson, who has shown the greatest creativity and patience in dealing with every aspect of the production of this book. I would like to acknowledge the work of Ms Mary Sanders, who copy-edited this volume with great diligence. And, I would especially like to acknowledge the guidance and assistance that Dr Françoise Wemelsfelder has generously shown in creating this volume, through her insight, thoughts and wisdom about animal behaviour and welfare, and her friendship. Dr Liz Williamson was an inspiration that one could 'act' positively to create changes for the welfare of animals and humans alike; she inspires through example, as does Dr Hannah Buchanan-Smith.

During the planning stages when this project was just an idea, Jean-Marie Stassin and his small coterie of creatures (Kilda, Samson, Nenni, Penther) living at Ash Cottage provided much support and encouragement, as did dear friends and colleagues – Emma Clery, Sanjoy Roy, Anita Roy, Jim Gardner and Gavin Baxter, Marion Smith, Sandra Gibb, Marion Dow, Betty Bruce, Mamie Prentice, Rosie Wilson, Dario Floreano, Lynn Hanley, Vickie Glidewell, Pratap Rughani, Paul Garber, Jim King and Christopher Klimowicz. I am also deeply grateful to Merelyn Dolins and Eva Trupin, who have great faith in me. Philip Trupin will always remain as an inspiration, especially when it comes to thinking about animals.

Finally, I would like to acknowledge my deepest thanks to the animals for whom this book is really written. Many animals, as individuals, have touched my life: as a token of the trusting relationships that have been bestowed on me and the entrance into their 'mysteries' that I have been allowed to observe, I wanted this book to exist. In my education, I have 'used' animals in studying them. This relationship has always left me uneasy: an animal is not a book of verse to be opened and shut at will. The animal exists beyond the classroom hours, laboratory work and research – it behaves, lives, dies. I was and am determined that students' education in future times will not be fraught with similar decisions based on supposedly mutually exclusive alternatives of whether to pass a course or fail by aligning themselves with certain beliefs about animals and the meaning of their lives. An education can promote both ethical and scientific thinking simultaneously; they are definitely not mutually exclusive concepts. Thus, this book emerged to assist students with their own beliefs about how animals should be treated in education, in research, and in our society, and this book emerged to assist those who teach courses on ethics and animal welfare. Those who love animals are students of animals' nature, and I hope that this book enlightens those outside of the structured educational establishments as well.

We obtained joy in creating this book, but the ultimate joy will occur when we no longer have the need to write books such as these, when animals as separate entities, in their separate existences, are acknowledged in their respectful place.

Part I
Attitudes to animals

1

A look back in the mirror: perspectives on animals and ethics

FRANCINE L. DOLINS

A bundle of grey and gold hair, three *Eulemur fulvus rufus*, the red-fronted brown lemur, were huddled on an angled branch at least 50 feet above where I sat. Most lemur species are highly endangered; I watched these *fulvus*, recognizing that their fate relied not on their survival by finding food and shelter when necessary, but on the fate of a conservation project whose success precariously balances between various interconnected programmes of education, health, research, eco-tourism, and foreign investment. This delicate balancing act attempts to keep itself upright amidst an onslaught of human needs, desires, and at times, greed.

Lemurs are endemic only in Madagascar. This makes it a special place for those who are concerned about the conservation and welfare of wild species of non-human primates. Yet, for those humans who inhabit this land where the forest is their primary resource for survival, their needs come into direct conflict with the lemurs, a westernized version of a valued 'flagship' species. The Malagasy who live in rural areas depend almost entirely on the land, what they can grow and what they can extract from it. In the cities, the people are dependent on those who have grown and extracted from the land. Their immediate survival needs depend on land and forest clearance, and are not always in parallel with those of the many species of lemurs. The lemurs' state of conservation, however, may be of little consequence when human poverty and sickness are more than apparent. This example, when the needs of human animals come into conflict with those of non-human animals, is just one type of relationship in which humans have come to find themselves; just one side of a multi-faceted issue reflecting the many possibilities of human relationships to non-human animals. There are many other ways that humans relate to animals, such as by keeping them as pets, subjects in experiments, objects of study or reverence, the basis for subsistence, or financial gain.

In a recently published book, *The Frankenstein Syndrome: Ethical and Social Issues in the Genetic Engineering of Animals*, the well-known ethical philosopher, Bernard Rollin (1995), made the important point that most people consider ethical issues to be 'dilemmas'. A dilemma, according to Rollin is 'a situation or question where there are two and only two choices or answers, both of which lead to unsatisfactory conclusions' (1995, p. 9). Luckily, most ethical issues are not dilemmas, but are matters of choice (Rollin, 1995). An important point made by Rollin (1995) is that, in dealing with most ethical issues, we are in the process of choosing from a selection of many possible choices. Moreover, any choice made, whether informed or not, will have impact. The purpose of this book, *Attitudes to Animals*, is to inform in order to provide a foundation from which to judge in making ethical choices regarding animals. Most importantly, this book should prompt the reader to question their current views, attitudes, and perspectives from which their long-held beliefs may have been derived.

The two threads of this narrative so far can be understood as the intertwined fates of animals and humans as a consequence of ethical choices. Underlying these choices is the nature of the human–animal relationship and how (understood only from the humans' point of view), this relationship developed. Human perspectives on the human–animal relationship reflect what has been learned, both spoken as well as unspoken attitudes and assumptions, from our families, society, media, education, and in our type of employment. For instance, in James Serpell's chapter, 'Sheep in wolves' clothing? Attitudes to animals among farmers and scientists', he discusses how agricultural workers employ distancing mechanisms in relationship to the animals they tend from birth until slaughter, as an answer to their ethical choice. Likewise, Serpell illustrates how scientists and technicians working with animals in laboratories typically use similar psychological devices, also compensating for the sacrifice of 'the many' by keeping a few animals as 'mascots'. This portrays the difficult emotional states of animal 'exploiters' in relation to the animals they tend, as well as their response to public criticism and pressure.

As a primatologist, I am often asked why I study primate behaviour and how the information I learn is useful for improving the lives of humans. Underlying these questions are certain implicit assumptions. The first is that non-human primates are not worthy of consideration in their own right; and the second assumption is that non-human animal life has no intrinsic value because the world was created for humans' use. These assumptions maintain a human-centred perspective. The origins of these beliefs, however, are not necessarily pervasive across all peoples, but rather arise from particular concepts present in particular cultures. In the chapter by Phyllis Passariello, 'Me and my totem: cross-cultural attitudes towards animals', perspectives on the human–animal relationship are examined through an anthropological looking glass. As examples of humans' perspectives towards animals and our shaping of this long relationship, Passariello discusses the domestication of animals and totemism. What characteristics have we idealized and shaped in generations of animals, and what characteristics of animals have we taken on, to wear or to hide behind, as disguises? As Passariello states, 'Attitudes, of course, do not fossilize', and so our study across cultures is a window on our personal and cultural evolution of attitudes towards animals: the tension between the cultural and the natural. It is, of course, the attitudes that we eschew that create the foundations for our choices.

Considering and even producing animals' existence merely for our own use suggests that we have taken on the role of 'creator', 'protector', or 'steward', tending and preserving nature for our own purposes. Control over animals is one consequence of our anthropocentric attitude toward animals. Humans have exerted their control over what was once wildlife, in zoos, sanctuaries, private menageries, and in laboratories for conservation, breeding, cloning, as well as for experimental purposes, respectively. Our goal has been, and is, to maintain this control in order to preserve endangered species. This is even true in the face of the likelihood that it was our control in the first place that has brought many species to the very edge of their existence. The problem that now faces us is how to keep endangered species from disappearing. The answer eludes us, as we can only but watch many species of animals and plants head toward extinction.

Do some humans have the right to maintain control over animals' lives in order to save them from other humans and/or from extinction? Charles Bergman's chapter 'In the absence of animals: power and impotence in our dealings with endangered animals' investigates this thorny issue of control over animal life, and where it may lead us and them. He examines the plight of the last wild California condors as an example. David Cooper's chapter 'Human sentiment and the future of wildlife' takes the argument of natural preservation from yet another perspective. Wildlife management exemplified our concern for animals' continuing presence on earth. Cooper discusses human concerns about the rights of animals to exist, highlighting the four reasons that dominate human concerns: the right of individual animals, the right of a species, the benefit to humans, and the benefit to the ecosystem. The chapter concludes that the final realistic (not idealistic) solution to the future of wildlife is to create and maintain 'relative' wildernesses which are inhabited by managed wild animals. The reader is called upon to view this solution not as second best, but as a pragmatic means which allows for human interactions with animals and for species survival. Under these circumstances Cooper considers this to be a limited interaction, although one which provides an enriching life experience. Mary Midgley's chapter 'Should we let them go?' takes yet another perspective on human control over animals. She discusses control in relation to the ethical problem of whether it is correct to keep animals in captivity at all, and if so, in what manner. Thus, taken together, these chapters provide various and at times contradictory views on the subject of human control over animal lives and destinies.

Some radical animal rights groups suggest that we should limit or cease our interaction with animals altogether, for animals' benefit. They question whether it is necessary for humans to interfere with animals at all in their daily lives. However, if we do limit these experiences, are we limiting our development as human beings as well? Patricia Hindley's chapter '"Minding animals": the role of animals in children's mental development' deals with the issues of human–animal interactions as it affects the development of a moral self. Correspondingly, Andrew Petto and Karla Russell, in their chapter entitled 'Humane education: the role of animal-based learning' explore the challenge of educating children humanely using concern for animals as the basis for ethical behaviour. They do not prescribe a set of rules but invite creative means by which to achieve a pathway to this goal. Petto and Russell define the aims in two ways. The first is to have the learning process provide the opportunity to understand the 'natural' and the 'cultural' animal. Secondly, they suggest modifying teaching methods with a view towards developing humane attitudes in children about the natural lives of animals, the effects of captivity, their role in the ecosystem, and resultant costs and benefits to both humans and animals. Similarly, David Dewhurst's chapter, 'Alternatives to using animals in education' is devoted to viable alternative methods for teaching biology to students while maintaining humane standards, ethical responsibility, and respect for animal life.

Through these chapters, I am reminded of a conversation in which a colleague bemoaned the fact that, once we begin the study of any system, we irrevocably alter that system. He explained that, once we impose measurements or observations even with careful regard, we cannot know with complete certainty the original state of that system. This, my colleague argued, was the irony of science: that we can never know what a system was like prior to our study of it (also see Mackenzie, 1997).

Similarly, our interactions with animals will invariably create changes in their lives, which may be for better or for worse. And, our study of animals will create changes in our human lives, which may ultimately prove to be an enriching process, or not, depending on our subjective perspectives and on the framework through which we perceive and define our experiences.

The scientific view of animals, from the influence of Behaviourism, is one of objectivism, ignoring the subjective qualities of animals altogether. However, this perspective often leads to animals being treated as 'black boxes', responding purely to environmental stimuli without reason or sentience. In this manner, the question of animal subjectivity affecting their actions and behaviour was not taken into account in scientific thought and methodology. Currently, scientific perspectives on animals are being re-examined. Henk Verhoog's chapter, 'Animals in scientific education and a reverence for life' carries us through logical argument, detailing the basis of scientific inquiry of animals' behaviour and the limitations of this methodology. In doing so, Verhoog makes suggestions for a paradigmatic shift in scientific inquiry. Similarly, Françoise Wemelsfelder in her chapter 'The problem of animal subjectivity and its consequences for the scientific measurement of animal suffering' describes a research model that uses animal subjectivity as a means to understand the motivation for animals' behavioural actions. 'Are animals aware of their own suffering?' is a fundamental question in animal welfare, and Wemelsfelder together with Verhoog provide a theoretical framework and experimental basis for investigating the subjective states of animals. It is only through research of animals' subjective states that this important question can be addressed.

The underlying subjective states of animals may also be investigated via other methods. Susan Healy and Martin Tovée present a chapter discussing the environmental effects on neurophysiology during development, 'Environmental enrichment and impoverishment: neurophysiological effects'. Their discussion centres on experimental evidence from animals raised in impoverished (versus enriched) environments where interaction with the environment and stimulation were minimal or highly controlled. The result is that, not only is the animal's ability to problem solve severely impaired, but its overall behavioural repertoire is significantly reduced. The behavioural deficits are reflected in developmental neurophysiological deficits as well. The authors point out that the major neurophysiological difference between an animal raised in an impoverished environment from that of an enriched environment is a significant decrease in cortical connections. Thus, the lack of variability in the behavioural repertoires of captive animals inhabiting impoverished environments can be correlated with, if not explained by, a lack of neuronal connections and density in important higher order processing areas of the brain. The benefit of this chapter in this book is to provide the reader with an understanding of the neuro-physiological mechanism(s) directing behavioural variability. This chapter also provides empirical evidence of the long-term effects of impoverished environments on animals, not only behaviourally, but also neurophysiologically. It enables us to have a greater understanding of how to alter the environments to provide stimulation for captive animals. Incidentally, information gained through this chapter may aid in establishing additional means to measure and enhance the welfare of animals on a neurophysiological plane in correlation with their behavioural repertoires, as well as provide benefits to diminish effects of disease and trauma in humans and animals.

Descartes declared that animals did not possess the capacity to process rational thoughts or to experience pain (1956). An extension of the Cartesian ideology set the past criteria for distinguishing humans from other animals, based on four major differences. Humans alone were thought to have the ability to use tools and language, to possess self-awareness and consciousness, and to be able to anticipate and plan intentionally for future events. In contrast to Descartes, Darwin (1859) believed that there were biological, psychological as well as emotional continuities between humans and all other animals. Conceptually, this placed all animals, including humans, along the same plane of subjective existence.

Following Darwinian tradition, the research of Jane Goodall (van Lawick-Goodall, 1976) and others (e.g. McGrew, 1992) have provided strong evidence for chimpanzees' abilities to both use and create tools. Additionally, language use in apes has been studied extensively by numerous researchers (e.g. Savage-Rumbaugh et al., 1986; Fouts, 1973; Patterson & Cohn, 1990). While the controversy still exists about whether apes generate novel sentences flexibly using syntactical structures (Terrace et al., 1979; Miles, 1994), recent evidence from studies of captive, language-reared pygmy chimpanzees (Pan paniscus) has provided strong support for their use of complex communication, their ability to use semantics in sign-based languages, and to comprehend syntax in human speech (Rumbaugh, 1996). Studies of the cognitive aspects of non-human animal behaviour complement both the study of language abilities and these animals' capacity for self-awareness and/or consciousness (see Parker, Mitchell & Boccia, 1994). The two remaining frontiers are the areas of self-awareness and consciousness, including empathy, and anticipation and intentionality. These are not wholly separate from each other: an animal that can anticipate a future can anticipate a future event where it is likely to be aware of its own suffering.

In Richard Byrne's chapter 'Primate cognition: evidence for the ethical treatment of primates' the themes discussed are non-human primate intelligence and capacity for anticipating future unpleasant events. Byrne provides evidence for classifying whether a non-human primate is sentient of its own future pain and thus worthy of 'welfare' considerations. This chapter should be read in relation to Françoise Wemelsfelder's ideas for comparison. Donald Broom's chapter 'Animal welfare: the concept of the issues' suggests that the welfare of an animal should be attended to in relation to its "needs, freedoms, happiness, coping, control, predictability, feelings, suffering, pain, anxiety, fear, boredom, stress and health". Broom provides a detailed conceptual and theoretical discussion of ways in which to assess the welfare of captive animals. On a related theme, in 'New perspectives on the design and management of captive animal environments', David Shepherdson's chapter is on the theoretical background and examples of environmental enrichment for captive animals in zoological parks. By modifying the environment in which captive animals live, their behavioural repertoire may be transformed from a paucity of stereotypical behaviours to a greater number of more natural types of behaviours that correspond to their wild counterparts.

It is obvious from the widely different backgrounds of the authors included in these chapters that opinions vary greatly amongst them. The authors of the chapters derive from interdisciplinary and varying backgrounds. Within these multiple viewpoints, the reader may choose lines of thought to follow on theoretical, conceptual and practical levels.

The section headed 'Animal awareness', a play on words, invites speculation on whether animals are self-aware, and our awareness of animals' behaviour and capacity for suffering. Robert Young's chapter 'The behavioural requirements of farm animals for psychological well-being and survival' addresses areas of concern in the welfare of captive animals, and in particular, agricultural animals. Given that the main purpose of farm animals is to provide financial benefit to their owner and food for people, the question of farm animal welfare is often only considered in relation to the monetary costs and/or economic effects. However, Young's chapter takes us beyond the financial concern of costs and benefits to the human owners and consumers, and deals with the underlying psychological states of these farm animals. Approved methods of animal husbandry for farm animals often deprive these animals of basic needs. The questions that

Young answers are whether farm animals have specific behavioural needs, what kind of suffering will be incurred if these needs are not met, and, what environmental modifications can be achieved to enhance these animals' behavioural repertoire? But, what if the behavioural needs of a captive animal are satisfied? Will the basic satisfaction of needs be enough to ensure an animal's 'happiness'? James King's chapter 'Personality and the happiness of the chimpanzee' deals with this question: where can we draw the line between when an animal has had its welfare addressed sufficiently so that we can claim that it exhibits 'psychological well-being', and when it is actually 'happy'? Also, are the two the same or can they be distinguished from each other? If we were to establish psychological well-being for humans, would we also be able to claim that they were happy? By taking a novel twist on the almost cliched term 'psychological well-being', King evaluates previous research in this area and presents findings, illuminating the contexts and personalities that lead to chimpanzees' states of happiness.

The intentions of the following chapters, as the titles of this book and introduction suggest, is for the reader to take a look in the mirror once more, asking, what is it to be human, what to be animal, and what are the nature of the relationships therein? The text should provide the reader with the means to glance back to past assumptions, to analyse, test and challenge both new and old perspectives on these issues by following the diversity of its many authors' thoughts. Finally, the subject matter of this book should pose a challenge to the reader to make choices, ethical choices, about their own human stance in their relationship to non-human animals, and by expansion to other human animals.

References

Darwin, C. (1859). *The Origin of Species*. New York: Hurst.

Descartes, R. (1956/1637). *Discourse on Method*. New York: Liberal Arts Press.

Fouts, R. S. (1973). Acquisition and testing of gestural signs in four young chimpanzees. *Science*, **180**, 973–80.

Lawick-Goodall, J. van (1976). Early tool use in chimpanzees. In *Play*, ed. J. Bruner, A. Jolly & K. Sylva, pp. 19–46. Cambridge: Cambridge University Press.

Mackenzie, D. (1997). Through the looking glass. *The Sciences*, May/June, 32–7.

McGrew, W. C. (1992). *Chimpanzee Material Culture: Implications for Human Evolution*. Cambridge: Cambridge University Press.

Miles, H. L. W. (1994). ME CHANTEK: The development of self-awareness in a signing orangutan. *Self-awareness in Animals and Humans: Developmental Perspectives*, ed. S. T. Parker, R. W. Mitchell & M. L. Boccia, pp. 254–72. Cambridge: Cambridge University Press.

Parker, S. T. , Mitchell, R. W. & Boccia, M. L. (1994). *Self-awareness in Animals and Humans: Developmental Perspectives*. Cambridge: Cambridge University Press.

Patterson, F. G. & Cohn, R. H. (1990). Language acquisition in a lowland gorilla: Koko's first ten years of vocabulary development. *Word*, **41**, 97–143.

Rollin, B. E. (1995). *The Frankenstein Syndrome: Ethical and Social Issues in the Genetic Engineering of Animals*. Cambridge: Cambridge University Press.

Rumbaugh, D. M. (1996). Beast machines of the monkey wars. *Contemporary Psychology*, **41**(4), 316–18.

Savage-Rumbaugh, S., McDonald, K., Sevcik, R. A., Hopkins, W. D. & Rubert, E. (1986). Spontaneous symbol acquisition and communicative use by pygmy chimpanzees (*Pan paniscus*). *Journal of Experimantal Psychology: General*, **115**(3), 211–35.

Terrace, H. S. , Petitto, L. , Sanders, R. & Bever, T. (1979). Can an ape create a sentence? *Science*, **206**, 809–902.

2

Me and my totem: cross-cultural attitudes towards animals

PHYLLIS PASSARIELLO

Human beings are part of the so-called animal kingdom. Universally, we humans seem to sense or intuit our proximity and affinities with other animals yet likewise require our distinct identity from other animals. Anthropology by definition is overtly anthropocentric and speciesist, relegating non-human animals to the category of 'other'. Ironically then, when looking cross-culturally, most humans to some extent appear to categorize certain animals, depending on culturally specific criteria, as allo-animals, as allegorical and/or mythological and totemic stand-ins for them(our)selves! Other creatures are a fact of life in our greater ecological niche(s), but other creatures also seem to hold a place in the universal human psychic 'niche' as well. In a culturally malleable, dynamic system, humans seem simultaneously to anthropomorphize and yet polarize other animals, recognizing both the close links and the dilemmas inherent in the intimacy. The dog can be best friend, or frightening assailant, or even dinner, largely depending upon cultural constructions, and depending on how the identity and power issues of the dog are specified by the humans involved. The tension of this self/other dynamic which we have towards other animals forms the core impulse behind all cross-cultural attitudes toward animals, no matter the superficial, specific, ethnographic variations in behaviours and feelings about animals.

In the ethnographic record, each culture has many ways in which animals interact with people, physically and metaphorically; likewise, each culture includes a variety of culturally specific attitudes toward specific other species. For example, foragers like the Ju/'hoansi of southern Africa (see Lee, 1994), or the Inuit of the Canadian Arctic (see Rasmussen, 1929; Wallace, 1966; Harris, 1988) are renowned for their pervasive animism, their recognition that all creatures and even things have 'souls'. Pastoral peoples like the Tuareg of north Africa (see Rodd, 1926) consider their camels their most precious 'possession' in the same way that the Saami of northern Europe value their reindeer and their reindeer-herding dogs; the Saami have even been noted to consider 'human-cum-dog' as *a* if not *the* basic relationship in their culture (see Anderson, 1986). The Maasai of east Africa (see Spencer, 1988) with their individually named and pampered cattle, or the Tsembaga Maring of New Guinea (see Rappaport, 1984) with their highly valued and also individualized pigs, or the horticultural and foraging Huaorani of Ecuador (see Kane, 1996) who hunt woolly monkeys but also keep them as pets, all respect and utilize non-human animals as an integrated part of each of their cultures. The peasant and largely vegetarian cultures of parts of India (see Harris, 1988) both religiously revere cattle as they maximally utilize them as 'resources'. In the post-modern cultures of the United States and England, a complex and in some ways anomalous cultural pastiche represents the human/non-human animal since, for example, pets provide major industries as do blood sports. Similarly, many non-human animals are bred as food for these heavily carnivorous groups, while at the same time, these cultures support many vegetarian, 'green', and animal activist subgroups as well.

With the larger anthropological and universalist perspective introduced above, two main topics are most relevant for exploring cross-cultural attitudes toward other animals: so-called animal domestication and totemism. Both topics have prehistoric, historic, diachronic and synchronic dimensions for this discussion.

Domestication

The very term 'domestication', of course, is political, and implies a power differential and intentionality that is blatantly contextual and cultural, an 'attitude' in itself. None the less, popular opinion today suggests that the domestication of animals (the human taming or 'producing', with concomitant genetic changes, of a new variety of animal from a wild variety which then is able to be bred in captivity and likewise unable to live in the wild) is a major characteristic along with the practice of agriculture, of the Neolithic Period, or New Stone Age, beginning about 10 000 years ago (Solecki, 1971). This is a conservative date. The earliest so-called animal domestication probably occurred about 16 000 to 12 000 years ago, first in the Middle East, and first with the dog (Trager, 1995; Grolier, 1995; Harris, 1988). However, according to some anthropologists, '...there is evidence from the sex and age distribution of prey animal bones that as early as 15 000 BP, pre-Neolithic Middle-Eastern hunter-collectors were exerting a considerable amount of control over the wild fauna in their habitat' (Harris, 1988; citing Russell, 1986, 71–5). Specifically, Trager (1995) cites fossil evidence from a cave near Kirkuk, Iraq, which indicates the presence of domesticated dogs. The remains were dated to about 16 000 BP by fluorine analysis, and have been interpreted as evidence of the domestication of dogs from Asian wolves, which were used in the Near East presumably for tracking game (Trager, 1995, 2). Dogs were followed in terms of domestication by goats, sheep, and eventually cattle and cattle-like animals in the ethnographic record in the Old World (see Harris, 1997; Trager, 1995). First domestication, of course, was not the last, and there is ample evidence from Star Carr in Yorkshire, for example, of domesticated dogs in the British Isles about 9300 BP to 9500 BP (Trager, 1995, 4; Harris, 1997, 133). Harris postulates that the changing ecology of Mesolithic times in Europe and elsewhere in the Old World, from the sweeping grasslands of the post-glaciation period which favoured large herd animals to the developing forests of the later period which favoured different sorts of animals in a more sheltered environment, necessitated different hunting skills and techniques, which included the evolution of the dog/human companion relationship. The dogs were invaluable, especially in the forest, for their tracking abilities (Harris, 1997, 133).

The differences between the first dates of human occupation and the differences in the quantity and varieties of indigenous fauna in the Old and New Worlds add other dimensions to the domestication issue. Most anthropologists accept an African genesis for proto-humans, and most anthropologists accept a much later arrival (11 000 BP to 65 000 BP, depending on one's acceptance of dating techniques and validity of certain contested sites) for modern humans in the Americas. Clearly, there were many more plant and animal resources for Old World peoples than for New World peoples, partially due to the greater land masses of the Old World, and partially due to the massive extinctions after the last ice ages. Diamond (1996) offers another view on the Old/New World contrasts. He suggests: 'Another reason for the higher local diversity of domesticated plants and animals in Eurasia than in the Americas is that Eurasia's main axis is east/west, whereas the main axis of the Americas is north/south' (Diamond, 1996). Consequently, species in the Old World, wild or not, could spread more easily across large latitudes without encountering massive climatic and other changes whereas in the Americas, species would quickly encounter vast changes in climate and day-lengths, travelling north/south (Diamond, 1996). In fact, in the New World, of course, only the Andean llama and alpaca, besides the dog and the tiny guinea pig, were ever domesticated, and never spread beyond the Andes; and similarly, only the Mesoamerican wild turkey was (semi-)domesticated and never spread from its initial site (Harris, 1997).

Attitudes, of course, do not fossilize, and we can never know exactly how earlier peoples felt about their animals. However, looking cross-culturally at the ethnographic record, and extrapolating back from attitudes apparent today, we can muse on the differences between relationships of people with different animals. For example, the dog as a hierarchical pack animal can be seen as a predictable candidate for human companion, or as a scavenger animal as a likely co-domesticate with humans, since the developing relationship in the Mesolithic period, for example, was mutually beneficial to both species, whereas with the sheep, as a counter-example, such an intimate mutuality seems less plausible – sacrificial lambs and rams, farmer jokes, and Scottish clones, aside! The cow, on the other hand presents a more complicated grey area, since there are cultures such as Hindu groups in India where the cow is religiously revered. Similarly, the Maasai and other groups in East Africa individually own and name cows, which form a major base of prestige in the culture. Likewise, in other cultures, tame pigs, elephants, or camels, and so on, may participate on different levels in a variety of co-relationships with humans, depending on the culture. Generalizations are difficult to make, except to say that most of the animals who exist culturally as allo-animals in one culture or another are mammals, though some birds certainly qualify, such as the falcons revered in medieval Europe and contemporary Saudi Arabia, or even the crocodile which has an allo-animal status in parts of Africa. To some extent, the labelling and treatment of an animal as an allo-animal, rather than standing as a cross-cultural, existential fact, depends on the anthropomorphizing categories particular to a culture. In the same way that Guthrie (1993) sees anthropomorphism as the basis of all religion, one can argue that our own human tendency to see only through our own eyes likewise operates in the construction of our culture, and thus our culturally based attitudes towards various animals. Anthropomorphism itself springs from the human condition, from our continual, inner and conscious vacillation about the so-called nature/culture or human/animal issue.

An interesting and related perspective on anthropomorphism and cross-cultural attitudes towards animals is presented by the historically documented interest cross-culturally, particularly in so-called developed, complex societies, in so-called feral children (see Candland, 1993). Canines as allo-animals are well represented by feral children stories in the ethnographic record. Wolf children, children supposedly raised by wolves, abound from ancient Roman times to 18th and 19th century Europe, to 20th century India, not to mention numerous traditional societies with wolf clans, totems, and coyote tricksters!

Likewise, related to cross-cultural interest and folk stories about feral children, and certainly related to the human inclination to try to tame wild things, is the idea and reality of the zoo. Of course, any zoo, no matter how progressive and innovative, is arguably a prison. M. Landzelius and K. M. Landzelius (unpublished observations) have pointed out how zoos are part of what they call 'the net of colonialism . . .', partaking in '. . .the legitimization of the cultural as natural . . .'. An analogue to the zoo which also illustrates the colonial roots of anthropology is presented by the displays of human beings at various carnivals, expositions, and world fairs, such as the Chicago Columbian Exposition of 1893 where there was a re-created Eskimo Village with real Inuit people on display, or the St Louis Exposition of 1904 where more than '. . .ten thousand indigenous people, from all over the world, lived and were exhibited in diorama-like displays. . .' (Mathe, 1996, 53). Zoos and human zoos illustrate the tendency in our culture to make so-called exotic animals and so-called primitive people operate analogously for us as tools to tame and appropriate the 'other'.

Looking at English legal history presents yet another viewpoint on attitudes towards other animals, and illuminates how this particular cultural viewpoint has had far-reaching ramifications for the so-called western, developed, 'First' world. In England, through the 12th century, animals were regarded as chattels. When King John signed the *Magna Carta* in 1215, he legalized an interesting distinction between wild and domestic animals in terms of property, 'wild' animals were owned by the King; 'domestic' animals, which were designated only by the generic title of 'cattle' and were established in terms of their usefulness, i.e. economic benefit, to humans, could be owned by individuals (Tannebaum, 1995, 559). These distinctions carried through into the laws of England, of the colonies, and ultimately, into the so-called common law of the United States of America (Tannebaum, 1995)! Other legal twists underline, though, the evident allo-animal tendencies of English and other early European laws. Records of several pre-Victorian cases describe situations where animals are equated with humans judicially, animals could be present in court and somehow 'testify' by their presence, and also animals could be tried and even hung for their crimes (Ritvo, 1990)! The pre-Victorian human profession of 'rat-rhyming' consisted of a man who attempted to communicate verbally with rats in an effort to persuade them to leave human premises (Ritvo, 1990). Today, some people subscribe to the idea of an 'earth antenna', in a quasi-scientific/psychic attempt to communicate with human-offending other creatures, such as ants (D. McMeekin, unpublished observations). The inclination to anthropomorphize other animals is strong.

A clear example of what can be framed as perhaps a co-evolution, socially for humans, biologically for the cattle, is the fascinating development of the Durham Ox in early 19th century England. Ritvo's study (1990) about what she terms the 'Barons of Beef', documents the experimental stock breeding of very large oxen in England by wealthy farmers. She directly relates the development of these 'elite cattle' to '. . .a persuasive rhetoric of self-assertion and display. . .'(Ritvo, 1990, 46,49). There is ample evidence which demonstrates, in her words, the connection between '. . .the ideal bovine and the social structure that it symbolically affirmed. . .' (Ritvo, 1990, 46). The breeding today of various types of show-animals from roosters to poodles, cross-culturally, can be placed in such a co-evolutionary, or in some cases, co-exploitative, context.

In the last 25 years, feminist theory and post-modernist studies generally have influenced anthropological thought in all areas, particularly questioning traditional bases of ethnographic authority and even pointing out the colonial roots and thus the hegemonic legacy of anthropology. A productive development of this critique is the appearance of a 'co-operative' paradigm arising along beside the traditional 'hierarchical' paradigm applied to every aspect of human studies from evolution to current gender roles. There are also studies of cooperation, ecologically and evolutionarily, among non-human animals as well (Dugatkin, 1997). For the study of so-called domestication, the co-operative paradigm leaves open new doors for interpreting how humans and other animals might have developed, co-evolved their/our inter-relationships. Gregory Bateson's decades of work on what he calls 'an ecology of mind' is an example of this type of thought (Bateson, 1972, 1988); development of the 'Gaia' hypothesis presents a similar avenue (Lovelock, 1979, 1988; Merchant, 1992), as does the development of the journal, *Co-evolutionary Quarterly* (renamed *Whole Earth Review*), and the rise of the so-called 'green' and 'animal rights' movements. The term co-domestication was coined in 1982 (Coppinger & Smith, 1982), radically redefining the parameters of inquiry on the issues, and introducing new perspectives such as how, '. . .some species [were] virtually volunteering for domestication. . .' (Anderson, 1986). What unifies these diverse trains of thought is the *de*-centring of humans from the various issues involved. Theoretically and historically, a burgeoning de-centring of human beings represents, of course, a major shift from the legacies of Aristotle, the Enlightenment, and even modernity. A simple study of domestication is no longer possible.

Totemism

Even before Sir Edward Tylor coined the word anthropology in 1871, J. F. McLennon coined the word totemism (in 1869–1870), and the concept was destined to shape the development of the discipline. The word totemism is derived from the Algonkian, specifically from the Ojibwa language from the word *ototeman*, which means, '. . .he is my relative' (Litz, 1979, 3551), but was seized upon by early anthropologists essentially to stake out a terrain for the field. McLennon's original definition stated that, '. . .tribes in the totem state believed themselves descended from or of the same breed, as some species of animal or plant, which was their "symbol and emblem", and "religiously regarded" or taboo". . .' (McLennon in Willis, 1994, 2); and J. G. Frazer said in 1910, 'A Totem is a class of material objects which a savage regards with superstitious respect, believing that there exists between him and every member of the class an intimate and altogether special relation. . . . Totemism is thus both a religious and social system' (Frazer in Willis, 1994, 2). Of course, this set the stage in 1912 for Emile Durkheim's classic discussion of aboriginal Australian totemism in *Elementary Forms of Religious Life*, where he suggested that totemism was an elementary form of religion and also indicated that what primitive people really worshipped was their own society. Despite Victorian ideals and social Darwinian prods, the ethnographic evidence did not corroborate the archaic religion connection for totemism. In fact, more recent investigations of Australian indigenous cultures continue to isolate the social organization and resultant strong and multiformed totemism in the Australian cases in point as unusual and atypical of many other foraging societies (Testart, 1988).

   Frank Boas, of course, the particularist and 'father of American anthropology', could be expected to have objections to any grand explanation of totemic behaviours. In fact, although there was no one position on totemism pushed by the Boas school, Boas's student Alexander Goldenweiser from 1910 to 1937 developed ideas on totemism, which were universalistic in nature and in fact foreshadowed (and, according to W. Shapiro (1991, 599), actually 'diffused' to. . .) Lévi-Strauss's structuralist interpretation of totemism in the 1960s. In 1911–12, at the same time as Durkheim was presenting his ideas, Goldenweiser put forth what he called 'The

Pattern Theory' where he generalized about the structural units in a totemic system. Later in 1918, he discussed the 'form and content' of totemism; to use Warren Shapiro's words: '…Goldenweiser has now liberated the notion of totemism from exclusive association with clan organization and extended it so as to include other forms of metaphorical connection between human and animal taxonomies' 1991, 304). Goldenweiser even included contemporary cultures in his attentions, noting that totemic inclinations crosscut time and place. In reference to 'phenomena' such as totemic classifications, Goldenweiser notes:

> Therefore we find that those phenomena are omnipresent in primitive society and also extend, in attenuated forms, into … modern culture, as witnessed by the … tendency to anthropomorphize the psychic life of animals, … or by the rich store of animalistic metaphor and allusion used in connection with human countenances, characters [and] affairs *(Goldenweiser (1918, 291–4) in W. Shapiro 1991, 604)*.

Years before Lévi-Strauss and *The Savage Mind* (1962), Goldenweiser convincingly universalizes the totemic impulse. None the less, in 1920, Van Gennep was still able to list 41 theories of totemism rampant in Europe alone (Willis, 1990, 4)!

Claude Levi-Strauss, the French structural anthropologist, of course, in the 1960s revitalized the totemism debate by emphasizing the universal, structural aspects of the cross-cultural trait. Lévi-Strauss developed the idea of animal symbolism as a universal human tendency to use the so-called natural world as a mirror for humans' cultural world; we order our cultures in this way because our minds are ordered in this way. Levi-Strauss likewise countered prevailing functionalist explanations of totemic behaviours, which suggested that particular totem animals were chosen, for example, because particular animals were crucial to the survival of particular cultures, with his famous remark that animals were not merely 'good to eat', but also, 'good to think' (Lévi-Strauss, 1963, noted in J. Shapiro, 1988). Generally, though, Lévi-Strauss's structuralism sought to decentre totemism and show that it was merely one of many traits which sprung from the universal categorizing impetus of human minds. He wanted to disarm the idea of totemism as an on-going anthropological discourse; totemism exists generally, universally, rather than as a specific, institutionalized correlate to a specific type of culture. Tiny Inuit animal carvings which serve as personal amulets, the animal guardian spirits which appear during Oglala Sioux vision quests, or the cartoon logo for the Chicago Bears are all totemic in impulse.

Neither Freud (1913) nor Marx (1968, 1859 – original) stayed out of the totemism debate. They both noted an interesting bit of McLennon's original definition of totemism as not simply 'the worship of animals and plants', but also the addition of behaviours of 'fetishism', in the sense of giving unreasonable reverence to the totemic objects. Rather than holy, there is something unholy and a bit pejorative about the concept of fetishism. Freud saw animal totems as displacements for the father-figure; Marx saw commodities as the unreasonable and unnatural totems of capitalistic society. In contemporary Mexico, Brandes (1984) has likewise noted that animal metaphors often play an important role in social control by permitting a displaced area of discourse for culturally 'disapproved or undesirable attributes' (Brandes, 1984, 207). Instead of the totems representing simply social reality for a group of humans, fetishism implies that the totems are in fact purposely *mis*representing social reality as 'natural' through the animal metaphors, for example, 'I'm a dog person' or, 'you're a cat person', are fetishistic disclaimers, perhaps, for justifying as 'natural' (and therefore out of our control) certain anti-social or at least questionable behaviours embodied by our allo-animal selves. The fetish, the symbolic totem, is a disguise, a cultural disguise, to deflect attention from the core misrepresentation of culture as nature. As Terence Turner elaborates:

The 'nature' incarnated in animal symbols is not simply the biological domain of animal species, adopted as a convenient metaphor for human social patterns. At the most fundamental level ... it consists of aspects of human society that are rendered inaccessible to social consciousness as a result of their incompatibility with the dominant framework of social relations ... These alienated aspects of the human (social) being ... are therefore mediated by symbols of an ostensibly asocial, or 'natural', character   *(Turner, 1985, 105).*

The animals, the totems are not simply a metaphorical mirror of culture, but rather a cultural disguise, a rationale, a deep justification for having the construct of culture at all. Maybe humans long to be 'natural' as well as cultural. And, of course, both states (or neither) are possible and true simultaneously.

Despite the discrediting of early claims in the totemism debate, some see a 'totemism revival' today in anthropology. Willis, for instance, points out that the surge in interest in ecology, especially human ecology since the 1970s has caused a re-examination of the nature/culture debate, and the place of humans. Willis notes that:

...the sense of interconnection between nature and culture, between human and animal ... which Victorian anthropology saw as a fascinating error of primitive man [sic] ... has now been rehabilitated in Western scholarly thought as an accurate reflection of existential reality ... Western culture, it seems, is now in a phase that might almost be called neototemistic *(Willis, 1994, 6)*.

He then goes on to talk about the 'fruitful tensions' between the principles of separation and continuity that comprise the What is human?/What is an animal? debate.

An interesting if polarized position on the human/animal debate is presented by the work of Spiegel (1996), who draws a direct comparison between human slavery and oppression of non-human animals by humans. She literally compares racism and speciesism, drawing analogy after analogy, particularly between the historical record of slavery in the United States and current non-human animal industrial practices in so-called developed nations. The polarity and vehemence of her argument relates to the anthropological literature about cannibalism, because of the de-humanizing requirements of both slavery and cannibalism. Anthropologists typologize cannibalism into two major categories: endocannibalism and exocannibalism (for example, see Dole, 1985; Clastres, 1985). In other words, in the ethnographic record, peoples who cannibalize at all most often seem either to eat their relatives or their enemies; one can eat a relative to preserve continuity and ancestral power, or eat an enemy to neutralize 'its' power. As with slavery and non-human animal oppression behaviours, power and identity issues are intimately involved with cannibalistic behaviours.

Now in the 'green' era, the era of animal advocacy, every child is aware of the category of endangered animal, every child knows of Jane Goodall and the plight of the chimps, our closest relatives, one of our now-noble allo-animals. From the chimps', the white rhinos', the African elephants', the Siberian tigers' points of view, maybe not only are people 'good to eat', but also, 'good to think'. Who's the totem? – Moby Dick or Ahab?? Are totemic animals finally enacting their revenge? For humans, of course, particular animals continue to be 'good to think', and maybe even, 'good to be'.

Turning to what might seem to be individualized, neototemism in the post-modern world, elephants and non-human primates, especially monkeys and apes, present convincing examples of animals which repeatedly appear as allo-animal, totemic choices in the ethnographic record. Elephants continue to attract humans because of their size, their exotic look, and their tractable nature. Monkeys and apes, as our simian look-alikes, continue to attract and repel because of their obvious *alter-ego* potential. Do these totems serve as mirrors or as scape-apes, so to speak, for us? Probably both, probably simultaneously, probably more.

From the time of prehistoric cave paintings of mammoths, to the worship of Ganesha in India, to Ringling Brothers circus attractions to contemporary lore about elephant graveyards, and contemporary researchers like Cynthia Moss, Parbati Barua, and Joyce Poole, who have devoted their lives to the study of the elephant, there is documentation of the co-relationships between humans and elephants. Likewise, with monkeys and apes, from Egyptian tomb paintings to Ajanta frescoes to Hanuman statues in Asia, to King Kong and Tarzan stories, to the contemporary 'Leakey's Angels' of anthropological fame, life researchers such as Jane Goodall for chimpanzees, Dian Fossey for gorillas, and Birute Galdikas for orangutans, there is ample and on-going documentation of the co-relationships between monkeys, apes, and humans. These two types of non-human animals are simply examples of the importance and strength of the totemic impulse, cross-culturally, and across time and spaces. An interesting aside here, an area for further inquiry, is the alignment of contemporary, 'First World' female researchers/advocates with particular allo-animals, perhaps indicating a new form of personal neototemism!

The shifting boundaries of the self vibrate for humans with our perceptions of the other. Cross-culturally, humans participate in a Hegelian dialectic of desire where, in fact, one's identity is articulated and felt as one's desires, often expressed with precise, cultural specificity. Paradoxically, humans experience simultaneously an individual and a social/cultural self; this paradox creates an ever-present and unrequited longing for resolution which, in turn, evokes and provokes for us across time, space and cultures, a variety of seemingly personal totems, icons, heroes, self-set challenges and touristic quests. The quest for the self is always the quest for the other; maybe the cultural self is the other for the individual self, part of the other within. For Willie Nelson, 'My heroes have always been cowboys', while for Jane Goodall, a chimp is as good as a saint. Somehow, they are each celebrating and thus appropriating their loneliness, fetishizing it, reformulating their own otherness into a virtue, or at least into a totem!

Does the totemic impulse reveal a metaphorical mirror or a deflective, cultural disguise? Do the totems embody social, even psychic solidarity, or rather help us to cope with the alienation of being existentially cultural? Are totems always cultural or can they be less (more?) than cultural? How does the totemic impulse frame the longing self, the raw desire, which is the nugget of discontent at the core of the human condition? How is this totemic impulse at the core of our attitudes towards other animals?

Conclusions

Human domestication claims and behaviours, and human totemic impulses and behaviours are at the heart, cross-culturally, of human attitudes toward other animals. Although these attitudes and behaviours take many cultural forms, they all seem to connect to the human dilemma, the mediation of nature and culture, the acceptance of our biocultural reality, the on-going condition of the longing self. In our increasingly self-conscious yet humanly de-centring world, perhaps we can continue to co-evolve a sustainable and ethically viable stance as human animals in the world in relation to the other animals here, and still satisfy what appear to be the universal psychic and physical needs of being human.

Acknowledgements

Special thanks for bibliographic assistance to Myrdene Anderson, Kyra Landzelius, Francine Dolins, Ann Silver, Mary Beth Garriott, Connie Klimke, Kristin Havill, Misty Schmitt, and Jane Nelson.

References

Anderson, M. (1986). From predator to pet: social relationships of the Saami reindeer-herding dog. *Central Issues in Anthropology* (A Journal of the Central States Anthropological Society), 6(2), 3–12.

Bateson, G. (1972). *Steps to an Ecology of Mind.* New York: Ballantine Books.

Bateson, G. (1988). *A Sacred Unity: Further Steps to an Ecology of Mind.* A Cornelia and Michael Bessie Book.

Boas, F. (1916). The origin of totemism. *American Anthropologist*, **18**, 319–26.

Brandes, S. (1984). Animal metaphors and social control in Tzintzuntzan. *Ethnology*, **23**(3), 207–15.

Candland, D. (1993). *Feral Children and Clever Animals: Reflections on Human Nature.* Oxford, UK: Oxford University Press.

Clastres, P. (1985). Guayaki cannibalism. In *Native South Americans: Ethnology of the Least Known Continent*, ed. P. Lyons, Prospect Heights, Illinois: Waveland Press.

Coppinger, L. & Smith, R. (1982). Livestock-guarding dogs that wear sheep's clothing. *Smithsonian*, **13**, 64–73.

Diamond, J. (1996). How domestication of animals proved easiest for Europeans. Faculty Research Lecture. UCLA. Http://www. bruin. ucla. edu/FRL/ Diamond/003. html

Dole, G. E. (1985). Endocannibalism among the Amahuaca Indians. In *Native South Americans: Ethnology of the Least Known Continent*, ed. P. Lyon, Prospect Heights, Illinois: Waveland Press.

Dugatkin, L. A. (1997). *Cooperation Among Animals: An Evolutionary Perspective.* Oxford Series in Ecology and Evolution. New York and Oxford: Oxford University Press.

Durkheim, E. (1912). *Les Formes Elementaires de la Vie Religieuse*. Paris: Presses Universitaires de France.

Frazer, J. G. (1910). *Totemism and Exogamy: A Treatise on Certain Early Forms of Religion and Society*. London: Macmillan.

Freud, S. (1913). *Totem and Tabu*. Leipzig and Vienna: Heller.

Goldenweiser, A. (1910). *Totemism, An Analytical Study*. Boston: American Folklore Society.

Goldenweiser, A. (1918). *Form and Content in Totemism*. Indianapolis: Bobbs-Merrill.

Grolier's Multimedia Encyclopedia (1995). Version 7. 04, Mindscape, Inc., IBM.

Guthrie, S. E. (1993). *Faces in the Clouds: A New Theory of Religion*. New York: Oxford University Press.

Harris, M. (1988). *Culture, People, Nature*. 5th edn, New York: Harper Collins.

Harris, M. (1997). *Culture, People, Nature*. 7th edn, Longman, An imprint of Addison Wesley Longman, Inc.

Kane, J. (1996). *Savages*. New York: Knopf Inc.

Lee, R. (1994). The Dobe Ju/'hoansi. Fort Worth, Texas: Harcourt Brace.

Lévi-Strauss, C. (1962). *The Savage Mind*. Chicago: University of Chicago Press.

Lévi-Strauss, C. (1962). *Totemism*. Chicago: University of Chicago Press.

Lévi-Strauss, C. (1963). *Structural Anthropology*. Chicago: University of Chicago Press.

Litz, R. J. (1979). Totemism. *Encyclopedic Dictionary of Religions*, 3551.

Lovelock, J. (1979). *Gaia: A New Look at Life on Earth*. New York: Oxford University Press.

Lovelock, J. (1988). *The Ages of Gaia: A Biography of Our Living Earth*. New York: W. W. Norton.

Marx, K. (1986, 1859 – original). *The Communist Manifesto*. New York: Modern Reader Paperback.

Mathe, B. (1996). *Jessie Tarbox Beals' Photographs for the 1904 Louisiana Purchase Exposition*. *VRA Bulletin* (Visual Resources Association) **23**(4), 53–61.

McLennan, J. F. (1869), (1870). The worship of animals and plants. *Fortnightly Review*, **6**, 4049–27, 5622–82; 7 (1870), 194–216.

Merchant, C. (1992). *Radical Ecology: The Search for a Livable World*. New York and London: Routledge.

Morels, E. (1995). *The Guinea Pig: Healing, Food, and Ritual in the Andes*. Tucson: University of Arizona Press.

Rappaport, R. (1984). *Pigs for the Ancestors: Ritual in the Ecology of a Papuan New Guinea People* 2nd edn. New Haven: Yale University Press.

Rasmussen, K. (1929). *The Intellectual Culture of the Iglulik Eskimos*. Report of the 5th Thule Expedition, 1921–1924, 7(1), Trans. W. Worster. Copenhagen: Glydendal.

Ritvo, H. (1990). *The Animal Estate: The English and Other Creatures in the Victorian Age*. London: Penguin.

Rodd, R. (1926). *The People of the Veil*. London: Macmillan.

Russell, K. (1986). *The Behavioral Ecology of Early Food Production in the Near East and North Africa*. London: BAR Series 391.

Shand, M. (1995). *Queen of the Elephants*. Mackays of Chatham: UK.

Shapiro, J. (1988). Gender totemism. In *Dialectics and Gender: Anthropological Approaches*, ed. R. Randolph, D. Schneider & May Diaz, Boulder: Westview.

Shapiro, W. (1991). Claude Levi-Strauss meets Alexander Goldenweiser: Boasian anthropology and the study of totemism. *American Anthropologist*, **93**(3), 599–610.

Solecki, R. (1971). *Shanidar: The First Flower People*. New York: Knopf Inc.

Spiegel, M. (1996). *The Dreaded Comparison: Human and Animal Slavery*. New York: Mirror Books, Institute for the Development of Earth Awareness.

Spencer, P. (1988). *The Maasai of Matapato*. Manchester, UK: Manchester University Press.

Tannebaum, J. (1995). *Animals and the Law: Property, Cruelty, Rights. Social Research*, **62**(3), 539–607.

Testart, A. (1988). Some major problems in the social anthropology of hunter–gatherers. *Current Anthropology*, **29**(1), 1–13.

Trager, J. (1995). *The Food Chronology: The Food Lovers Compendium of Events and Anecdotes, from Prehistory to the Present*. NY: Henry Holt and Co.

Turner, T. (1985). Animal symbolism, totemism, and the structure of myth. In *Animal Myths and Metaphors in S. America*, ed. Gary Urton, pp. 49–106, University of Utah Press.

Wallace, A. F. C. (1966). *Religion: An Anthropological View*. New York: Random House.

Willis, R. (ed). (1994). *Signifying Animals: Human Meaning in the Natural World*. Routledge.

3

Sheep in wolves' clothing? Attitudes to animals among farmers and scientists

JAMES A. SERPELL

Introduction and methods

Within the last 30 years, western attitudes to non-human animals and their treatment have undergone some revolutionary changes (see Midgley, 1994; and also see Chapter 11, this volume). Methods of exploiting animals that used to be accepted without question are now criticized on ethical grounds by growing numbers of people, and animal-related issues which were once considered the domain of small, idealistic minorities have become the subject of widespread public and political controversy. This remarkable change in moral emphasis is obliging certain sectors of the animal-using community to re-examine and justify their attitudes and activities towards animals in the face of mounting public criticism. This chapter describes how the members of two distinct animal-using groups – livestock farmers and research scientists – are responding to this kind of pressure.

In popular humane and animal rights literature, animal 'exploiters' are commonly portrayed as being simply callous or cruel. Such simplistic characterizations tend to reinforce feelings of dislike and distrust, while at the same time discouraging constructive dialogue between those on opposite sides of the debate. Indirectly, they therefore also constitute a barrier to improvements in animal welfare. The aim of the present study was to move beyond these stereotypic depictions to explore the actual attitude constructs employed by people involved in various consumptive, animal-using activities.

The material presented in this chapter was collected *in situ* (i.e. on farms and in laboratories) in the form of anonymous, recorded interviews with individual volunteer subjects. A small fraction of this material was then subsequently used in the production of a radio documentary.[1] Subjects

---

[1] 'Killers with a Conscience' BBC Radio 4, 28 October 1989 (producer: Miles Barton).

were asked about the animals they used and the practical aspects of their husbandry and treatment, and how they themselves felt about employing animals for such purposes. They were also invited to counter or respond to some of the typical arguments levelled against their use of animals by supporters of animal protection or animal rights. Interviews were open ended and were only discontinued when subjects considered that they had adequately expressed their attitudes and values concerning animals (see also Serpell, 1989). Only a small number of individuals were interviewed and, as volunteers, all of them were self-selected. However, since they were informed at the outset that the material they provided would be unattributed, there is little reason to assume that their statements and comments were biased or unrepresentative.

In the time since this study was completed, the results of a number of other more exhaustive and detailed attitude surveys have been published. These have explored attitudes among research scientists (Arluke, 1988; Gluck & Kubaki, 1991) and animal rights supporters (Plous, 1991; Herzog, 1993; Paul, 1995), but have not yet included analyses of livestock farmers' attitudes.

Results

Farmers

All of the farmers interviewed were somewhat defensive, and most of them showed a marked tendency to shift or deflect the blame for modern farming methods away from themselves. Critics of intensive farming were often accused of being out of touch with economic realities, and most farmers claimed that the only reason they reared animals intensively was because consumers 'demanded' cheap meat, eggs or dairy products. By implication, then, most of them would have preferred to raise animals more extensively, if only it were economically viable. In short, the consumer is ultimately to blame. Few of the farmers interviewed slaughtered their own animals, even for home consumption, and they therefore did not feel entirely responsible for their demise. Indeed, some specifically avoided inquiring too deeply into the fate of the animals once they left the farm. As the owner of a large egg production unit put it: 'I think they get turned into meat pies, but frankly I'd rather not know what happens to them.'

Without exception, farmers placed considerable emphasis on the health and productivity of their charges, as if freedom from disease, rapid growth, and high reproductive performance were entirely synonymous with good welfare. Many also insisted that their animals were actually quite 'happy', and that they would be less happy – cold, uncomfortable, vulnerable to disease, more inclined to fight, etc. – if kept under more extensive conditions. They were also quick to point out that farm animals are 'better off' than wild animals, in the sense of being safer and more comfortable, and that they would not exist at all if not for farming. In this way, many farmers tended to cast themselves in the role of 'good shepherd', i.e. as protective and custodial agents, rather than as purely exploitative ones. A number also regarded themselves as guardians of the countryside and of rural traditions, of which livestock farming was viewed as an integral part. A few also argued that meat-eating, and hence meat production, is a part of 'nature' and the natural relationship between a predator and its prey. As agents of this process, these individuals saw themselves as serving the common interests of meat-eating society as a whole.

The farmers in this survey made certain efforts not to get to know their animals too well as individuals. On the larger farms this appeared to be relatively easy since it is impossible to get to know thousands of superficially similar animals individually. At this level of detachment, the animals, like sausages, can be abstracted to the status of mere units of production. The process is often reinforced by the use of euphemisms. In fur farming, for example, mink or silver foxes are typically referred to as a 'crop' which will eventually be 'harvested', rather than as animals which will be slaughtered and flayed. On more traditional farms and small-holdings, distancing devices (see Serpell, 1986) of this kind were less easy to maintain, and other, more subtle techniques were employed. Some farmers avoided naming the animals that were destined for slaughter (names, after all, signify personal status) or they were given special, often comical, names such as 'Mint Sauce' or 'Rashers', as constant verbal reminders of their ultimate fate.

Division of labour within the farming community undoubtedly helps to dilute the burden of individual responsibility. Most larger farms employ stockmen whose job it is to look after the animals and, as far as possible, get to know them and recognize when an individual is not doing well. These stockmen, however, play little part in determining the animals' eventual fate. Executive or management decisions of this type are usually made by the farm owner or manager who, conversely, has relatively little day-to-day contact with the animals.

Scientists

Similar attitudes were prevalent among scientists engaged in biomedical and behavioural research involving animals. Like farmers, many claimed that they were only doing what society demanded but, in addition, bio-medical researchers were also able to argue that their activities might actually improve health or save lives. Science therefore tended to be perceived by its practioners as a nobler and more self-righteous pursuit than farming. Most of the scientists interviewed were unwilling to claim that the pursuit of knowledge, on its own, was sufficient justification for using animals in research. Off the record, however, many drew attention to the difficulty of predicting the potential beneficial outcomes of basic, as opposed to strictly applied research.

In general, animals that were used for relatively harmless, non-invasive procedures were often named and treated like pets. This was particularly true of primates, dogs, cats and other larger mammals, especially if they were tame and co-operative. Conversely, scientists seemed to make con-scious efforts not to become too attached to, or familiar with, animals that were destined for more invasive research, and smaller, less easily personi-fied species, such as rodents, were more frequently used. According to Arluke (1988), this process of detachment; this objectification of the ani-mal, is achieved in various ways. On grant applications, animals fall into the category of 'consumables'; they are often implicitly or explicitly de-fined as tools, instruments or sources of data, and they are commonly identified by means of numerical codes rather than names. The person conducting the experiment may also be reluctant to come into contact with a conscious animal, preferring to have it delivered either dead or already anaesthetized. Newcomers into a laboratory, such as research students and new technicians, may initially treat the experimental animals with affection and sympathy. But, often this is strongly discouraged by more experienced colleagues.

Among my informants, it was not considered unusual for one or two individual animals to be kept aside as named mascots or pets. By nurturing and personifying these selected individuals, it appeared that these scientists were in some way attempting to atone or compensate for their treatment of the animal's less fortunate, and more anonymous, fellows. Again, Arluke (1988) cites several examples of this type of behaviour. In one case, a group of laboratory technicians conspired to save the life of a rabbit, nicknamed 'Fat Cheeks', who was destined to be exsanguinated as part of an experi-ment. This involved covertly removing and storing blood from the animal over several days in advance, and feeding the rabbit up again between bleedings. Fat Cheeks was then falsely logged as being killed, along with the other rabbits, and secretly smuggled out of the lab. One individual I interviewed harked back to the days when stray dogs could be used for research, and admitted that the knowledge that these animals were doomed anyway eased his conscience considerably. Arluke, on the contrary, found that most of his informants preferred the notion of purpose-bred laboratory animals – that is, animals whose sole function in life was to serve as experimental material. This difference between the two studies is probably cultural. In the United States, where Arluke's study was conducted, horror stories involving the discovery of stray or stolen pets being used for experiments are still commonplace and not entirely without substance (Clifton, 1992). In contrast, legislation in the United Kingdom has now rendered the use of non-purpose-bred dogs or cats for research virtually impossible.

All of the scientists in this survey emphasized how much they disliked the business of killing animals, and a surprising number were vegetarians. Euphemistic terminology was also widely used. Animals, for example, tended to be 'sacrificed' rather than killed in research. Arluke (1988) attaches considerable significance to the term sacrifice and it forms the main theme of his paper. He points out that, in a religious context, the sacrificial victim embodies the whole community. It is the object of identification, respect and even deification. In the realm of science, the experimental animal is also frequently represented as a substitute for a human being, a substitute that ultimately gives its life in order to further the interests and survival of the community. A simpler interpretation might be that the term sacrifice functions purely as a means of disguising or deodorizing the true nature of the act being performed (see Serpell, 1986).

As in farming, division of labour among researchers appears to be important. The scientists who design the projects and determine the experiments, often have little if any involvement in the care of the animals. He or she thus avoids getting to know them too well as individuals. Conversely, the technicians who look after the animals are not generally responsible for deciding their fate. Such divisions sometimes gave rise to a certain amount of job dissatisfaction among animal care technicians, particularly when the value of the research or the justification for killing the animals is not properly explained.

Although necessarily superficial, this brief account of attitudes to animals among farmers and scientists reveals a number of consistent themes. Above all, it suggests that farmers and researchers are genuinely ambivalent about their roles as exploiters of animal life, particularly in the present climate of rapidly changing public opinion. This ambivalence has not prevented these individuals from continuing to pursue their chosen occupations, but it does appear to call for the adoption of a consistent pattern of attitudes and values, some of which appear designed to deflect criticism, shift blame or expiate guilt. The strategies commonly adopted include:

(a) denying or diluting responsibility, either via division of labour and hierarchical chains of command, or by 'blaming' consumers;
(b) filtering out or misrepresenting evidence of animal suffering;
(c) expiating guilt through compensatory acts of benevolence toward selected individual animals;
(d) maintaining a *distance* from the animal, e.g. avoidance of naming, or any other behaviour that might lead to the personification or sympathetic identification with the animal;
(e) disguising the fate of animals through the use of euphemistic terminology, e.g. 'harvesting' pelts, 'sacrificing' experimental animals, and so on;
(f) perceiving oneself as the agent of some higher moral purpose, e.g. saving lives, serving human interests, preserving the countryside, and so on.

Discussion

In the field of social psychology, it is widely recognized that humans tend to apportion their social and moral obligations according to how 'close' or similar others are to themselves (Deaux & Wrightsman, 1984). Thus people are more inclined to behave altruistically towards those who are familiar or related, for example, friends, neighbours, kinsmen, etc., and are proportionately less inclined to treat these individuals in harmful ways. Conversely, people tend to feel less inhibited about harming more distant categories of individual, such as strangers or foreigners. At the same time, an individual's perceived social distance is rarely immutable. As a result of social contact, interaction and observation, there is an inevitable tendency for distant categories of individuals to drift closer to oneself. Familiarity, in this case, breeds sympathy rather than contempt. And, one obvious implication of this process is that, if Jack ultimately intends to harm Jill, he must either have some way of absolving himself of responsibility for his actions, or he must take steps to prevent Jill becoming too familiar in the first place.

The results of the present survey suggest that similar rules apply to animals, although the criteria people use for deciding whether an animal is close or distant are evidently more complex. Feelings of affinity with animals, for example, may be based on superficial resemblances or 'cuteness' (e.g. giant pandas), actual biological resemblances (e.g. chimpanzees), apparent intelligence (e.g. cetaceans); recognition of similar feelings, motivations and needs (itself a product of familiarity), rewarding social relationships (e.g. pets), and various admirable qualities, such as beauty or physical prowess, with which people identify or wish they possessed (Burghardt & Herzog, 1989; Kellert, 1989; Serpell, 1990). Any, or all, of these criteria may qualify animals for quasi-human status, thereby creating a need for ways of distancing them and/or legitimizing their harmful treatment. As Rothschild (1986) points out: 'just as we have to depersonalize human opponents in wartime in order to kill them with indifference, so we have to create a void between ourselves and the animals on which we inflict pain and misery for profit.' Judging from the views expressed in this survey, farmers and scientists confront this problem on a regular basis, and respond to it in the predicted way: first, by denying sole responsibility for their actions towards the animals in their care, and secondly, by maintaining a certain physical, emotional or conceptual distance from them. In addition, acts of kindness towards selected individual animals may help to expiate a certain amount of guilt, and it is clearly important for both groups to see themselves as agents of the common good.

These findings may also contain some lessons for those seeking improvements in the treatment of animals in society. Confronted by the resistance of the animal-using establishment, supporters of animal protection have a strong tendency to resort to strident and simplistic accusations of cruelty, complacency or greed. Although occasionally justified, such allegations are often misplaced and frequently counter-productive. Irrespective of the moral rights or wrongs of animal exploitation, it is normal and appropriate for people to strive for consistency between their attitudes and their behaviour, and to resist attempts to contradict or undermine these constructs, particularly when their livelihoods are at stake (Festinger, 1957). It would therefore be unreasonable to expect farmers, scientists or any animal-using group to abandon their carefully constructed attitudes and values overnight or without a struggle. Furthermore, many of those who participated in the present survey were brought up and trained at a time when less critical views on animal use prevailed, added to which the fact remains that the bulk of their professional activities are still tacitly endorsed by a majority of the general public. The superior or sanctimonious moral tone commonly adopted by members of the animal protection movement is not only unlikely to prove effective in changing such attitudes, it is in danger of producing the obverse effect: of polarizing the debate still further, and driving animal-using groups into even more entrenched and oppositional positions. In the light of these considerations, campaigners on behalf of animals may achieve greater success by tailoring their arguments and strategies sympathetically to take account of the unavoidable professional and psychological difficulties confronted by those whose behaviour they seek to change.

References

Arluke, A. (1988). Sacrificial symbolism in animal experimentation: object or pet? *Anthrozoös*, **2**(2), 98–117.

Burghardt, G. M. & Herzog, H. A. (1989). Animals, evolution and ethics. In *Perceptions of Animals in American Culture*, ed. R. J. Hoage, pp. 129–51. Washington DC: Smithsonian Institution Press.

Clifton, M. (1992). Pet theft: can we stop it? *The Animals' Agenda*, April 13–18.

Deaux, K. & Wrightsman, L. S. (1984). *Social Psychology in the 80s*, 4th edn. Monterey, CA: Brooks/Cole Publishing.

Festinger, L. A. (1957). *A Theory of Cognitive Dissonance*. Stanford, CA: Stanford University Press.

Gluck, J. P. & Kubaki, S. R. (1991). Animals in biomedical research: the undermining effect of the rhetoric of the besieged. *Ethics and Behavior*, **1**, 157–73.

Herzog, H. A. (1993). 'The movement is my life:' the psychology of animal rights activism. *Journal of Social Issues*, **49**, 103–19.

Kellert, S. R. (1989). Perceptions of animals in America. In *Perceptions of Animals in American Culture*, ed. R. J. Hoage, pp. 5–24. Washington DC: Smithsonian Institution Press.

Midgley, M. (1994). Bridge-building at last. *Animals and Human Society: Changing Perspectives*, ed. A. Manning & J. A. Serpell, pp. 188–94. London & New York: Routledge.

Paul, E. S. (1995). Us and them: scientists' and animal rights campaigners' views of the animal experimentation debate. *Society and Animals*, **3**, 1–21.

Plous, S. (1991). An attitude survey of animal rights activists. *Psychological Science*, **2**, 194–6.

Rothschild, M. (1986). *Animals and Man*. Oxford: Clarendon Press.

Serpell, J. A. (1986). *In the Company of Animals*. Oxford: Basil Blackwell.

Serpell, J. A. (1989). Attitudes to animals. *The Status of Animals*, ed. M. Palmer & D. Paterson, pp. 162–6. Wallingford: CAB International.

Serpell, J. A. (1990). All the King's horses. *Anthrozoös*, **3**, 223–5.

Part II

Animal awareness

4

The problem of animal subjectivity and its consequences for the scientific measurement of animal suffering

FRANÇOISE WEMELSFELDER

Introduction

In modern, intensive housing systems, animals develop a wide variety of abnormal behaviour patterns. The general public has become increasingly aware of this. With growing urgency, people wonder what it must be like to be a hen in a battery cage, or a chimpanzee in a laboratory testing-pen. This formulation, to ask 'what it feels like' to be a certain animal, is how members of the general public perceive and formulate the problem of animal welfare. In day-to-day, common-sense interaction with animals, we use a type of language which is inherently subjectivistic, psychological in character; that is, a language which indicates that we share the world with active, independent individuals who have their own, personal perspective on that world, and accordingly, their own needs, feelings and thoughts. A technical philosophical term for such a kind of language, such a level of discourse, is 'first-person-perspective' (Nagel, 1986). To describe behaviour in first-person-perspective terms is to assess what that behaviour means from the other individual's perspective. For example, one may say an individual likes to do this, or wants to do that. Generally, animals are accepted into that level of discourse, assuming that they too have their particular point of view. We say the cat wants to get out of the house, or the pig expects to be fed. And, upon witnessing an animal's abnormal, highly repetitive behaviour, we presume that the animal finds its situation unmanageable and highly distressing.

The task of scientists working in the field of animal welfare is to investigate whether the public's perception of animal suffering can be supported with objective facts. Does an animal's abnormal behaviour indicate real suffering, or is the abnormality only apparent to us, the human observers? Industry and government demand that, given the enormous economic interests, a neutral, unbiased scientific approach to this question is taken. Thus, scientists in the field of animal welfare must establish whether animals are indeed in some manner subjectively aware of their own situation, or whether we just project our human values and feelings upon them. It must be asked what animal awareness is, and how it can be measured.

However, this task immediately confronts us with a serious problem. Current standards for objective measurement *a priori* deprive an animal of its status as a subject, as a being with its own point of view. These standards are set to transform an animal into a physical object, a mechanical system. That is, they are set, in philosophical terms, to assess an animal's behaviour from a 'third-person-perspective' (Nagel, 1986). Accordingly, they will approach an animal's subjective experience as an 'object' for research in the standard physical, mechanistic sense of the word. This is where the problem lies. To approach animal suffering as if it were an object, in third-person perspective terms, is not the same as asking, in first-person-perspective terms, what it is like to be a certain animal in a certain situation. The question is whether it makes sense at all to approach the question of animal suffering in a mechanistic framework. It may well be that in regarding the animal as a complex mechanical system, current standards of scientific measurement lose the question as originally posed by the public.

One may consider this problem trivial, arguing that the scientist's task is indeed to get away from the public's subjectivistic level of discourse, and leave it behind for the sake of a detached, objective opinion. However, in such an approach, there is serious danger of throwing the baby out with the bath water. Is it possible to meaningfully assess questions of subjective well-being in animals, when the very level of discourse which has given rise to these questions is rejected? Would this not misconstrue the essential nature of subjective phenomena, defining these phenomena out of existence? Rather than rejecting common-sense perception of animals as subjects out of hand, perhaps science should seriously consider its possible validity, and seek to provide it with a reliable foundation. It seems essential that public and scientific understanding of the natural world eventually coincide, achieving agreement on an appropriate course of action, so as to alleviate potentially severe forms of animal suffering.

The aim of this chapter is to provide a general discussion as to how different conceptions of subjective awareness in animals affect our scientific understanding of animal suffering. The first section will provide a short overview of the different forms of abnormal behaviour which animals develop when housed in small, barren cages. Thereafter follows a

Fig. 4.1

discussion of two different models of animal consciousness and their consequences for the interpretation of abnormal behaviour. The chapter concludes by presenting suggestions for some starting points for the scientific measurement of chronic suffering in captive animals.

Abnormal behaviour in captive animals

Animals housed in small, barren cages generally interact less with their environment as time goes by. This is expressed in a variety of behaviours. The animals lie down and sleep more, and spend significantly more time sitting, often in rigid, seemingly unnatural postures (Buchenauer, 1981; see Fig. 4.1). On the other hand, they may over-react to novel and/or unexpected events with fearful and aggressive responses (Broom, 1986; see also Chapters 6 and 9 in this volume). Furthermore, captive animals may develop stereotyped patterns of behaviour. Such behaviour is highly repetitive and uniform; it is as if the animal has developed a compulsive habit. Laboratory primates, for example, frequently show stereotyped pacing, jumping and somersaulting (Fox, 1986).

As the animal spends more time in its cage, stereotyped behaviours tend to become increasingly directed towards the animal's own body. Primates may spend long periods of time masturbating, rocking their head and body, or eating and regurgitating their own faeces. They may also physically attack their own body in a highly aggressive way (Chamove & Anderson, 1981). Tethered sows may spend a long time chewing air, with no other apparent effect than the production of large amounts of saliva. Eventually, such self-directed behaviour may develop into compulsive self-mutilation. Laboratory monkeys gnaw at their own limbs and genitals, creating deep wounds, while parrots will pull out their feathers until completely naked (Morris, 1964). In summary, the overall decrease in interaction shown by captive animals becomes apparent in, first, a decrease in the variability of behaviour and, secondly, an increase in the self-directedness of behaviour (cf. Dantzer, 1986).

The interpretation of abnormal behaviour

Behavioural scientists mostly refer to the behaviour described in the previous paragraphs as 'abnormal' because, by and large, it does not occur in the wild. Moreover, it bears strong resemblance to behavioural pathologies in psychiatric patients and autistic children. In such human beings, it is accepted as self-evident that abnormal, disturbed behaviour is an expression of disturbed subjective well-being. In relation to animals, however, such an interpretation is not accepted as easily. It very much depends on whether or not we grant the various species any form of conscious awareness. This section will discuss two different models of animal consciousness, and their consequences for the interpretation of abnormal behaviour. First, the traditional view as currently accepted by most behavioural scientists is discussed, and it is argued that its conception of animal consciousness as a purely internal, intellectual process is too limited. Then a more dynamic view of conscious awareness is presented, which regards behaviour as an inherently expressive process.

The traditional view: conscious awareness as intellectual reflection

Most behavioural scientists think of the behaviour–consciousness relationship in the dualistic framework typical of western philosophical tradition, as inspired by the 17th-century French philosopher René Descartes. Descartes conceived of behaviour as an external phenomenon, induced and regulated by external stimulation, and therefore of mechanistic, or machine-like, nature. Consciousness, in contrast, he conceived as an internal, mental phenomenon, defined as 'thinking', 'intelligence' or 'introspection'. Whereas Descartes assumed that consciousness was a God-given attribute, scientists currently postulate that it is generated within the brain, on a 'higher' level of organization. Yet, like Descartes, they believe that conscious awareness does not come directly to expression in behaviour but guides it from 'within'. Thus, dualistic models character-istically 'split' behaviour into internal, perceptive processes on the one hand, and external, motoric processes on the other hand.

It is important to realize that the word 'internal' in this context has not just a spatial meaning. Nobody believes that upon opening the brain, or any other part of the body, conscious awareness will be found in a particular physical location. By using the term 'internal' a concept more fundamental than internal anatomical location is indicated; it means that conscious processes are fundamentally inaccessible to scientific measure-ment. Because behaviour is thought to be externally driven (i.e. mechanis-tic) and exclusively motoric, its scientific measurement is not regarded as problematic. Consciousness, however, is defined as non-mechanistic and non-motoric, so therefore it cannot be directly related to behaviour, and cannot be measured. Consciousness may actually exist in any number of forms, but the problem is that there is, according to this view, no direct way to test for its existence or measure its presence. As a consequence, conscious awareness has come to be regarded as an elusive phenomenon which cannot be observed through standard methods of scientific measure-ment; it acquires the status of a 'ghost-in-the-machine' (cf. Ryle, 1949).

The inevitable consequence of such a dualistic, 'ghost-in-the-machine' conception of conscious awareness is that it can never be proven beyond all reasonable doubt that animals are conscious, intelligent and emotional living beings. It is true that the recently revived interest in animal con-sciousness has led to many fascinating discoveries of intelligent behaviour in a large variety of animal species (e.g. Griffin, 1992; Dawkins, 1993). Animals can, for example, create ingenious tools to find food, they can communicate with complex signals almost resembling language, and they live in complex social groups and form relationships for life (Beck, 1980; Johnson & Norris, 1986; Cheney & Seyfarth, 1990). These facts have impressed the general public as well as scientists in various disciplines, and have led to a general upsurge in our moral concern for animals. Both Griffin (1992) and Dawkins (1993) argue that the study of intelligent behaviour allows us the best possible access to conscious processes in animals, and should, despite its indirectness, be given respectable scientific status.

However, the scientific community at large seems reluctant to accept apparently intelligent behaviour as an indication of animal consciousness. It is argued that the seemingly clever behaviour may well be the result of complex behavioural conditioning, induced by external factors in the animal's environment. That the behaviour appears clever does not necessarily mean that the animal is clever, that it truly 'understands', in the internal, reflective sense of the word, what it is doing (e.g. Epstein & Koerner, 1986; Kennedy, 1992). The same line of argument applies to the interpretation of abnormal behaviour. Such behaviour can also be regarded as a conditioned response to a barren environment; there seems no need to postulate that the animal 'understands' the abnormal character of its behaviour and suffers as a consequence (e.g. Dantzer, 1986; see also Wemelsfelder, 1993, for a discussion of various examples). Thus, within a dualistic model of the behaviour–consciousness relationship, it is simply not possible to establish reliable, hard criteria for the study of either animal consciousness or animal suffering.

It is important to realize, however, that this lack of criteria is not an empirical, given state of affairs. It is a direct consequence of a philosophical assumption, of the way in which the behaviour–consciousness relationship is conceived. As outlined in the introduction, standard methods of measurement conceive of animals primarily as objects, whose behaviour consists of various levels of interconnected mechanisms. Inevitably, this leaves conscious awareness, a non-mechanistic phenomenon, hanging out on a limb. Little else can be done, given mechanistic standards, than declaring consciousness to be a 'private' object. Consciousness thus acquires the status of an afterthought, an epiphenomenon which is causally irrelevant. I suggest, however, that this irrelevance, the so-called privateness of conscious awareness, is a direct consequence of the misguided attempt to understand consciousness within an essentially mechanistic, third-person perspective. Consciousness is not one kind of (non-physical) object, causing other (physical) kinds of objects. Instead, the concept of consciousness, as it functions in common-sense interaction with animals, denotes that animals are not mere objects, but subjects; that is, it indicates that a level of behavioural organization is perceived which requires a non-mechanistic, subject-related, first person perspective level of explanation. The question is whether such a conception of subjective awareness in animals is open to systematic and reliable investigation, so that it can be provided with a scientific foundation.

To consider this question, it is necessary to put the basic tenet of mechanistic models of animal behaviour to closer scrutiny. To assume that behaviour is a mechanical and externally regulated process is to assume that it is essentially reactive, or, in other words, that the causation of behaviour is essentially passive. Such reactivity can be conceived in two ways: as directly apparent, in simple, reflex-like stimulus–response mechanisms, or as indirectly apparent, where external forces of natural selection have, in the course of evolution, generated an internal system of pre-programmed rules. If such rules are very complex, they may be regarded as evidence of cognition, in which case we prefer to speak of internal 'representations' rather than rules. Yet, the principle remains the same. With regard to both reflex-like and cognitive processes, it is assumed that behavioural causation is an exclusively automatic, reactive process (McFarland, 1989). The question, however, is whether this assumption is correct. The alternative hypothesis is that the causation of behaviour is not merely reactive, but a predominantly active and self-generated process. Such a view has important consequences for the understanding of animal suffering, in that it would endow behaviour with a directly expressive, psychological character.

Towards a more dynamic view: conscious awareness as behavioural expression

The assumption that animal behaviour is of exclusively mechanistic nature may have led to a spectacular increase in our knowledge of an animal's physiology, but that does not imply that it provides a satisfactory explanation of all aspects of animal behaviour. The process through which animals engage in interaction with their environment in fact appears to be of an active, self-generated nature. Animals continuously pay active attention to their environment, scanning it for novel or unexpected information (Fig. 4.2). Also, they seek new opportunities for interaction through exploration and play. Such forms of active attention enable animals to develop new behavioural strategies in previously unencountered situations, endowing their behaviour with a variable and flexible nature (cf. Tolman, 1948; Fagen, 1982). Animals are not passively moved about by external events in their environment, they actively decide whether, and how, they will respond to these events. Of course, external factors and pre-programmed rules guide and influence behaviour, yet they do not cause it. In its dynamic, attentive aspects, behaviour is primarily voluntary and self-generated. Thus, animals are not merely reactive objects; they are subjects, active agents within their environment (for a more detailed discussion of this argument, see Wemelsfelder 1997a, and Wemelsfelder & Birke, 1997).

This principle leads to an alternative approach to animal consciousness,

Fig. 4.2

which is not dualistic in the traditional, Cartesian sense of the word. To recognize that the causation of behaviour is essentially voluntary (i.e. that animals are subjects), is to recognize that behaviour itself is a psychological, expressive process. Conscious awareness does not drive the behavioural machine from 'within', but comes to expression in action, as an emergent property of the behaving animal as a whole. This suggests that the way in which an animal pays attention to, and interacts with, a given environment directly expresses its subjective experience and awareness of that environment. In all the numerous subtle ways in which an animal orients its eyes, ears, nose, whiskers, or its whole body to a certain event, it expresses the way in which it perceives and evaluates that event.

From this perspective, to say that an animal is aware of its situation does not mean that it thinks 'intellectually' about that situation. Mind and body, perception and movement are not split, they form an integrated whole. Behaviour is an inherently expressive and not a purely mechanical, process. For example, we may say that an animal is curious (Fig. 4.3), or that it wants to get into the house (Fig. 4.4). Such descriptions are not per definition anthropomorphic and unscientific; they are qualifications of overt patterns of active attention, and as such they are, in principle, liable

Fig. 4.3

to operational definition and systematic investigation (see Wemelsfelder 1997*b* for elaboration on this point). This is in contrast to dualistic models, where the term conscious awareness strictly implies 'thought' which is separate from, and may be unrelated in origin to, behavioural processes.

A frequently made objection to such an integrative view of animal behaviour is that mechanistic explanations of behaviour are fundamentally more 'simple' and parsimonious (i.e. more 'objective') than other kinds of explanation. It is argued that it should therefore, in principle, be possible to give a mechanistic account of the active attentional aspects of behaviour, making a psychological account redundant. However active and spontaneous an animal's behaviour may appear, it can be accounted for by a set of more or less complex pre-programmed rules (e.g. Kennedy, 1992).

In response to these objections, there are two main points to be raised. First of all, whether or not a mechanistic account of active attention in animals can be given is in fact irrelevant. Such an account would not *explain* the active aspects of behaviour; it would merely force these aspects into a reactive framework. One does not *explain* the form of a circle by saying it is a round square; these are incompatible categories (Ryle, 1949). The active, self-generated nature of behaviour requires no further explana-

Fig. 4.4

tion; it is itself a causal principle, warranting the use of a psychological, first-person terminology. Active attention is not just another behaviour pattern, it is a mode of behavioural causation which cannot be explained in terms of another mode of behavioural causation. If this principle is ignored, considerable conceptual confusion may arise.

For example, models of neural organization tend to depend on the postulation of monitoring devices such as 'attentional filters', 'comparators' and 'general information processors'. It can be assumed that the postulation of such devices successfully explains the active nature of behaviour in terms of 'underlying' mechanisms, that it is 'really' these mechanisms (attentional filters, etc.) which 'cause' attentional behaviour (e.g. Sayre, 1986, plus peer commentaries). However, taking concepts which originate from a description of the actively behaving organism as a whole (it is the behaving animal that we observe monitoring its environment and making decisions), and then to apply these concepts to mechanical sub-elements, does not lead to a mechanistic explanation of that behaviour. It merely identifies the behavioural function of that particular mechanism (Bindra, 1984; Searle, 1990). The use of active metaphors in mechanistic schemes has, of course, proven to be extremely instructive. It does not, however, reflect a successful reduction of attentional concepts into a mechanistic scheme. On the contrary, it shows that our understanding of neural and genetic mechanisms is, and remains, dependent upon active, psychological qualifications of an animal's interaction with its environment.

Secondly, it is not clear what justifies the assumption that mechanistic explanations are generally more simple and parsimonious than psychological ones. Occam's razor or Morgan's canon are often mentioned in this context. However, as Rollin (1989) has pointed out, this seems misguided. Occam's razor says 'Do not multiply entities unnecessarily' (cited in Rollin, 1989, p. 75) and refers to explanatory units within any given level of explanation, not to the postulation of different levels of explanation. Morgan's canon suggests that we should not 'interpret an action as the outcome of the exercise of a higher psychical faculty, if it can be interpreted as the exercise of one which stands lower in the psychological scale' (cited in Rollin, 1989, p. 75). This statement obviously accepts a psychological level of explanation in general, but objects, like Occam's razor, to unnecessarily complex explanations within that level. Thus, there appears to be no general rule dictating that mechanistic explanations *per se* are more parsimonious than other kinds of explanations.

It follows, then, that first-person, psychological descriptions of animal behaviour cannot be replaced by third-person, mechanistic ones. Such descriptions are not based on indirect projection, as dualistic models assume, but refer to real and visible aspects of behaviour. They reflect the active, voluntary nature of behavioural causation and as such fulfil an indispensable role in the explanation of behaviour, on every level of organization. More specifically, psychological descriptions qualify overt patterns of active attention, and as such are, in principle, liable to operational definition and systematic investigation.

How can such a model of the behaviour–consciousness relationship facilitate the interpretation of the abnormal behaviour patterns which animals develop under intensive housing conditions? Two propositions can be made on the basis of this model. First, it may be proposed that the development of abnormal behaviour reflects the (gradual) disintegration of the voluntary and flexible character of behaviour. In the development of stereotyped behaviour patterns, for example, behaviour appears to become increasingly repetitive and fixated in form and function, and appears to lose its diversity and versatility (cf. Wood-Gush, Stolba & Baker, 1983; Dantzer, 1986). The increasing self-directedness of these patterns indicates that animals close themselves off from their environment, rather than interact with it. This cut-off can most dramatically be observed when animals living in a small cage are transferred to a larger, more enriched cage. These animals often do not respond to their new environment, but persist in old and rigid patterns of behaviour (e.g. Meyer-Holzapfel, 1968).

Secondly, it may be proposed that such an apparent process of disintegration is directly expressive of inner subjective disintegration, of chronic suffering. When first brought into its restrictive environment, the animal is likely to show normal levels of active attention. Due to a persistent lack of opportunities for normal species-specific interaction, however, the animal's attention may become more and more dispersed and dominated by inappropriate stimuli, such as the bars of its cage, or its own limbs. The animal 'does not know what to do', so to speak, and becomes increasingly listless and withdrawn. This stage may be characterized as boredom. As time goes on, the animal's attention may further disintegrate. The animal may become obsessively fixated upon its own body or other environmental features, as apparent in various stereotyped behaviours. It may also become apathetic, or over-react to unexpected events. Most animals show all of these symptoms at different occasions; together, they indicate that, rather than actively involved with its environment, the animal now has become helpless and cut-off. This stage may be regarded as depression and/or anxiety. Thus, as active attention breaks down, we see the animal losing its grip psychologically, becoming bored and then depressed (for a detailed discussion of this model, see Wemelsfelder, 1993).

Such a model of animal suffering does not imply that animals have an intellectual understanding of their abnormal behaviour and deprived situation. Apathetic animals do not necessarily sit in their corner reflecting upon their misery; as explained above, the proposed model does not take a dualistic split between behaviour and mental processes as a starting point. What the model implies is that animals emotionally experience their situation of enforced and chronic passivity as meaningless and highly distressing, and that this feeling comes to direct expression in the rigid and withdrawn character of their behaviour. That such animals are suffering is an observable phenomenon; in a non-dualistic approach, suffering is not internal or hidden, it is overt and publicly accessible.

The crucial question is of course whether such a model of animal suffering can be supported by scientific evidence. In other words, can the proposed view of conscious awareness in animals lead to a reliable scientific methodology for the measurement of animal suffering? The last section of this chapter will provide some starting points for dealing with these questions.

Starting points for the scientific measurement of animal suffering

Testing the proposed model of animal suffering has two aspects: investigation of changes in voluntary attention, and the interpretation of these changes as expressions of suffering. First, the hypothesis that impoverished housing conditions lead to a disintegration of voluntary attention can best be investigated by offering animals some kind of interactive situation. One could, for example, present a variety of novel objects to animals housed either in enriched or in impoverished conditions, and compare how these animals interact with those objects. If this were done at subsequent points over a prolonged period of time, one would predict that the animals' active attention for the novel situations would gradually decline, becoming increasingly erratic and passive as time went on. This prediction was tested at the Scottish Agricultural College in a study with 26 female growing pigs. In this study, we investigated the pigs' interaction both with the home pen environment, and with a novel test situation and a novel object. In the home pen, the most salient result was that pigs housed in impoverished conditions showed a significantly reduced diversity of interactive patterns compared to pigs housed under more enriched conditions (Haskell *et al.*, 1996). The results of the novelty tests reflected those found in the home pen; pigs kept in impoverished conditions responded to the novel situation and object with a significantly reduced diversity of interactive and manipulative behaviours compared to more enriched pigs (Wemelsfelder *et al.*, unpublished observations). This effect was present from the beginning of the experiment onwards, and suggests that impoverished housing conditions may bring about some change in the organization of behavioural attention. The question is what the nature of this change is; to investigate whether as predicted, it reflects a disintegration of voluntary attention, more research is needed.

However, the experiment also brought to our awareness a fundamental problem concerning the second aspect of the proposed model of animal suffering, i.e. the interpretation of observed changes in voluntary attention as expressions of suffering. We realized that the data of our study could never provide a basis for such interpretation, because of their fundamentally physical and quantitative nature. Standard methods of measurement simply are not designed to gauge the expressive, psychological aspects of behaviour, and their results therefore leave these aspects untouched. One may argue theoretically that expressive aspects are available for observation, but if one does not use an appropriate method of measurement, these aspects will continue to be highly elusive to scientific reasoning. In other words, if one wishes to develop a model of animal suffering, it is essential to reconsider dualistic conceptions of the behaviour–consciousness relationship not just on an abstract theoretical level, but on a practical level of measurement as well. Standard methods of behavioural measurement (i.e. ethograms) categorize behaviour in terms of discrete, mutually exclusive states of physical movement (e.g. walk, sit, sniff, play). Such categories however leave undescribed the dynamic transition between these states, that is the way in which an animal performs these physical movements.

What makes behaviour dynamic and expressive is that it takes place in time; to categorize the dynamic nature of behaviour is to describe its transitory, non-fixated character. To complement the static categories of standard ethograms, therefore, it is necessary to develop transitional categories. This is exactly how categories of active attention should be understood. Terms such as curious, bored, enthusiastic or timid do not refer so much to attentional states, but rather to attentional style; that is to the way in which an animal pays attention to its surroundings through active orientation of its body, limbs and sensory organs.

Such transitional categories of attentional style are not liable to standard procedures of quantification. It is not possible to score an animal's attentional style from moment to moment, as one would score categories of physical movement. An animal is not curious one moment and timid the next: attentional style emerges as time proceeds, and it is necessary to observe behaviour over a certain period of time before the style in which it is performed becomes apparent. This makes the categorization of attentional style an essentially qualitative process. Psychological categories such as curiosity and fear do not denote one particular state, but summarize and integrate the emergent, transitory attentional properties of an animal's behaviour. It is clear that such a process of qualitative categorization demands a more active role of the human observer than is the case in quantitative ethograms. But, rather than immediately dismissing this active role as anthropomorphic and unscientific, we should seek to understand and formalize it. Why could a qualitative method of categorization, a 'qualitative' ethogram, not have as much inter-observer validity as a quantitative method? It may follow different rules of measurement, but it is not therefore necessarily less reliable.

Starting points for formal qualitative ethograms can already be found in various scientific disciplines. For example, various categories of attentional style are used in rating scales designed to classify different forms of human depression (e.g. Parker *et al.*, 1990). Even though such rating scales define levels of depression partly on the basis of verbal self-report, they may provide a useful starting point for the measurement of animal depression. For example, the Emotions Profile Index, also based on human self-report, has already proven to be valid in animal studies (Plutchik & Kellerman, 1974). Qualitative rating scales have furthermore been applied successfully in the study of personality traits in cats and various primates (Buirski *et al.*, 1973; Feaver, Mendl & Bateson, 1976; Buirski, Plutchik & Kellerman, 1978; Stevenson-Hinde, Stillwell-Barnes & Zunz, 1980; see also King, this volume). Mendl and Harcourt (1988) note that personality traits such as 'curious' and 'active' refer to complex, emergent patterns of behaviour which remain elusive within standard recording methods. Recording methods designed to deal with such emergent patterns (e.g. Stevenson-Hinde & Zunz, 1978) could well be adapted for the measurement of animal suffering along the lines proposed here. Thus, expertise gained in various areas of research can be used as a basis for the development of a reliable, qualitative methodology for the measurement of chronic suffering in animals. A research programme to this effect is, at present, being instigated at the Scottish Agricultural College in Edinburgh.

In conclusion, our daily experience of animals as subjects may be provided with a scientific basis, through appropriate definition, categorization and measurement of psychological concepts such as boredom and depression. A qualitative methodology for the measurement of behavioural expression will not define questions of subjective well-being and suffering in animals out of existence, but will allow us to investigate these questions on their own ground. This, in turn, will allow us to submit public perceptions of animal well-being and suffering to scientific scrutiny in the most direct way possible.

References

Beck, B. B. (1980). *Animal Tool Behaviour: The Use and Manufacture of Tools by Animals*. New York: Garland STPM Press.
Bindra, D. (1984). Cognition, its origin and future in psychology. In *Annals of Theoretical Psychology*, ed. J. R. Royce & L. P. Mos, pp. 1–29. New York: Plenum Press.
Broom, D. M. (1986). Stereotypies and responsiveness as welfare-indicators in stall-housed sows. *Animal Production*, **42**, 438–9.
Buchenauer, D. (1981). Parameters for assessing welfare, ethological criteria. In *The Welfare of Pigs*, ed. W. Sybesma, pp. 75–95. The Hague: Martinus Nijhoff Publishers.
Buirski, P. , Kellerman, H. , Plutchik, R. & Weininger, R. (1973). A field study of emotions, dominance, and social behavior in a group of baboons (*Papio anubis*). *Primates*, **14**, 67–78.
Buirski, P. , Plutchik, R. & Kellerman, H. (1978). Sex differences, dominance and

personality in the chimpanzee. *Animal Behaviour*, **26**, 123–9.

Chamove, A. S. & Anderson, J. R. (1981). Self-aggression, stereotypy and self-injurious behaviour in man and monkeys. *Current Psychological Reviews*, **1**, 245–56.

Cheney, D. L. & Seyfarth, R. M. (1990). *How Monkeys See the World*. Chicago: University of Chicago Press.

Dantzer, R. (1986). Behavioural, physiological and functional aspects of stereotyped behaviour: a review and a re-interpretation. *The Journal of Animal Science*, **62**, 1776–86.

Dawkins, M. S. (1993). *Through our Eyes Only? The Search for Animal Consciousness*. Oxford: W. H. Freeman.

Epstein, R. & Koerner, J. (1986). The self-concept and other daemons. In *Psychological Perspectives on the Self*, ed. J. Suls & A. G. Greenwald, vol. 3, Hillsdale: Lawrence Erlbaum.

Fagen, R. (1982). Evolutionary issues in the development of behavioral flexibility. In *Perspectives in Ethology 5*, ed. P. P. G. Bateson & P. H. Klopfer, pp. 365–83. New York: Plenum Press.

Feaver, J. , Mendl, M. & Bateson, P. (1986). A method for rating the individual distinctiveness of domestic cats. *Animal Behaviour*, **34**, 1016–25.

Fox, M. W. (1986). *Laboratory Animal Husbandry; Ethology, Welfare and Environmental Variables*. Albany: State University of New York Press.

Griffin, D. R. (1992). *Animal Minds*. Chicago: University of Chicago Press.

Haskell, M. , Wemelsfelder, F. , Mendl, M. T. , Calvert, S. & Lawrence, A. B. (1996). The effect of substrate-enriched and substrate-impoverished environments on the diversity of behaviour in pigs. *Behaviour*, **133**, 741–61.

Johnson, C. M. & Norris, K. S. (1986). Delphinid social organization and social behavior. In *Dolphin Cognition and Behavior: A Comparative Approach*, ed. R. J. Schusterman, J. A. Thomas and F. G. Wood. Hillsdale: Lawrence Erlbaum.

Kennedy, J. S. (1992). *The New Anthropomorphism*. Cambridge: Cambridge University Press.

Lynch, M. E. (1988). Sacrifice and transformation of the animal body into a scientific object: laboratory culture and ritual practice in the neurosciences. *Social Studies of Science*, **18**, 265–89.

McFarland, D. J. (1989). *Problems of Animal Behaviour*. Harlow: Longman Scientific and Technical.

Mendl, M. & Harcourt, R. (1988). Individuality in the domestic cat. In *The Domestic Cat: The Biology of its Behaviour*, ed. D. C. Turner & P. P. G. Bateson, pp. 41–54. Cambridge: Cambridge University Press.

Meyer-Holzapfel, M. (1968). Abnormal behaviour in zoo animals. In *Abnormal Behaviour in Animals*, ed. M. W. Fox. Philadelphia: W. B. Saunders.

Morris, D. (1964). The response of animals to a restricted environment. *Symposia of the Zoological Society London*, **13**, 99–118.

Nagel, T. (1986). *The View from Nowhere*. Oxford: Oxford University Press.

Parker, G. , Hadzi-Pavlovic, D. , Boyce, P. , Wilhelm, K. , Brodaty, H. , Mitchell, P. , Hickie, I. & Eyers, K. (1990). Classifying depression by mental state signs. *British Journal of Psychiatry*, **157**, 55–65.

Plutchik, R. & Kellerman, H. (1974). *Emotions Profile Index*. Los Angeles, California: Western Psychological Services.

Rollin, B. E. (1989). *The Unheeded Cry. Animal Consciousness, Animal Pain and Science*. Oxford: Oxford University Press.

Ryle, G. (1949). *The Concept of Mind.* Harmondsworth: Penguin books.

Sayre, K. M. (1986). Intentionality and information processing: an alternative model for cognitive science. *Behavioral and Brain Sciences*, **9**, 121–66.

Searle, J. R. (1990). Consciousness, explanatory inversion, and cognitive science. *Behavioral and Brain Sciences*, **13**, 585–642.

Stevenson-Hinde, J. & Zunz, M. (1978). Subjective assessment of individual rhesus-monkeys. *Primates*, **19**, 473–82.

Stevenson-Hinde, J. , Stillwell-Barnes, R. & Zunz, M. (1980). Subjective assessment of rhesus monkeys over four successive years. *Primates*, **21**(1), 66–82.

Tolman, E. C. (1948). Cognitive maps in rats and men. *Psychological Review*, **55**, 189–208.

Wemelsfelder, F. (1993). The concept of boredom and its relationship to stereotyped behaviour. In *Stereotypic Behaviour: Fundamentals and Applications to Animal Welfare*, ed. A. B. Lawrence & J. Rushen, pp. 65–97, Wallingford: CAB-International.

Wemelsfelder, F. (1997*a*). The scientific validity of subjective concepts in models of animal welfare. *Applied Animal Behaviour Science*, **53**, 75–88.

Wemelsfelder, F. (1997*b*). Investigating the animal's point of view; an inquiry into a subject-based method of measurement in the field of animal welfare. In *Animal Consciousness and Animal Ethics*, ed. M. Dol, S. Kasanmoentalib, S. Lijmbach, E. Rivas & R. van den Bos, pp. 73–89, Assen: van Gorcum.

Wemelsfelder, F. & Birke, L. I. A. (1997). Environmental Challenge. In *Animal Welfare*, ed. M. C. Appleby & B. O. Hughes, pp. 35–47, Wallingford: CAB International.

Wieder, D. L. (1980). Behavioristic operationalism and the life-world: chimpanzees and chimpanzee researchers in face-to-face interaction. *Sociological Inquiry*, **50**, 75–103.

Wood-Gush, D. G. M. , Stolba, A. & Miller, C. (1983). Exploration in farm animals and animal husbandry. In *Exploration in Animals and Humans*, ed. J. Archer & L. I. A. Birke, pp. 198–210, London: Van Nostrand Reinhold.

5

Environmental enrichment and impoverishment: neurophysiological effects

SUSAN D. HEALY and MARTIN J. TOVÉE

The idea that the degree of complexity in an animal's environment may cause neurophysiological changes is not a new one. Nor, indeed, is the idea that performance of certain learning and memory tasks by an individual might, in some way, 'exercise' particular regions of the brain and that this 'exercise' might lead to changes in size or capacity of those brain regions. Most of us at some time have probably wished that our intellectual power could be enhanced by mind exercises in a way not too dissimilar from going jogging or working out in a gym. Equally, we might worry that failure to exercise our intellect might lead to it growing flabby and atrophying. Evidence does exist that the complexity of our environment has a strong effect on the development and maintenance of cortical connections. For most animals, the laboratory environment is usually far simpler and less stimulating than their natural environment would be. This sensory impoverishment may have a significantly detrimental effect on an animal's brain organization, particularly during its development. The aim of this chapter is to survey what is known of the effects of environmental complexity on neurophysiological development and to determine the key elements of the environment important for normal development.

The first experiments testing the neurophysiological effects of environmental conditions were done over 300 years ago. By the middle of this century, increased knowledge of brain development had resulted in a firm belief that brain size did not change in response to experience: the size and structure of the adult brain was purely and simply a product of development and developmental processes were genetically fixed. Decay as a result of ageing was the only cause for adult brain changes, excepting damage due to disease or injury. A reversal in this doctrine has been occurring over the last 30 years with the accumulation of evidence demonstrating a causal relationship between neural form and function and environmental experience (for review see Renner & Rosenzweig, 1987; Bedi & Bhide, 1988).

In this chapter we focus on two specific brain systems and examine in detail the outcome of environmental experience on these systems. The first is a sensory system, the visual system. There is a substantial body of work, particularly with mammals, on the neurophysiological changes brought about in the visual system by environmental effects. The second system we shall examine is the hippocampus. Perhaps surprisingly, there is much less evidence for such effects on this second system. The hippocampus is regarded as a higher-order processing system and plays a vital role in learning and memory. It might be expected, then, that the hippocampus would be an area showing significant changes in response either to an impoverished or an enriched environment. We discuss both the recent data and reasons why there may be such marked differences in the impact of environmental experience between different neural systems.

A sensory system: vision

The development of the visual system

The development of the visual system in the period following birth may be under the control of either genetic or environmental factors. Genetic control means that the structure and organization of the visual system is predetermined and independent of external factors. Partial or complete control by environmental factors would mean that the organization of the visual system could be altered by altering the input into the visual system. Input into the visual system can be examined in several ways: (a) by taking advantage of the natural variation in perceptual ability of some species (such as the variation in the number of retinal photopigments in New World monkeys (e.g. Tovée, 1994) or the abnormal organization of the retinal projections to the lateral geniculate nucleus in Siamese cats (e.g. Hubel & Wiesel, 1971), (b) binocular or monocular deprivation, (c) misalignment of the images in the two eyes, or, (d) by rearing animals in an environment with strictly controlled visual stimuli.

The normal arrangement of the visual system

In the normal adult monkey or cat, the retinal ganglion cells project to the dorsal lateral geniculate nuclei (LGN). These nuclei, located on either side of the brain, consist of six layers of cells. Each layer receives input from only one eye: layers 2, 3 and 5 from the ipsilateral eye, and layers 1, 4 and 6 from the contralateral eye. The topographic arrangement of the ganglion receptive fields is maintained in the LGN, so that each layer contains a complete map of the retina. The LGN projects primarily to the striate cortex or primary visual area, which consists of six principal layers (and several sub-layers) arranged in bands parallel to the surface. The axons from the LGN terminate on cortical neurons in layer 4. The neurons in layer 4 of striate cortex have centre-surround characteristics like those of the LGN and retinal ganglion neurons from which they receive input. However, neurons in other layers of the striate cortex show other properties. Most are responsive to a line of a particular orientation, and their response falls off when the orientation alters by more than 10 degrees. Some orientation neurons have receptive fields organized in an opponent fashion. Hubel and Wiesel referred to them as simple cells. For example, a line of a particular orientation might excite the cell if placed in the centre of its receptive field, but inhibit if moved away from the centre. Another type of neuron, called complex cells, have larger receptive fields and, although they respond to lines of a particular orientation, have no inhibitory surround. Many of the cells respond best to a line moving perpendicular to its angle of preferred orientation. Hubel and Wiesel also reported the presence of hypercomplex cells. These cells respond to the same stimuli as complex cells, except that their response is significantly reduced or abolished if the stimulus line extends outside their receptive field. The cells' response is end-stopped. The terms simple, complex and hypercomplex, suggest a hierarchy of feature detection we now know to be incorrect.

Although many neurons in striate cortex do respond to lines and bars, the optimal stimulus is a sine wave grating (De Valois, Albrecht & Thorell, 1978). Any waveform, no matter how complex, could be shown to be the sum of a series of sine waves of different frequencies and amplitudes. The dissection of the complex wave into its elementary components is called a Fourier analysis. Striate cortex seems to perform a sort of Fourier analysis, each cells detecting the presence of cycles of variations in brightness at a particular spatial frequency and at a particular orientation. The neurons which Hubel and Wiesel classified as simple and complex both respond to sine wave gratings, but in different ways. The simple cells are phase-sensitive, whereas complex cells are phase-insensitive. Hypercomplex cells are no longer considered to be a separate category, but a form of complex cell.

Cells in the retina, LGN and striate cortex of newborn, visually naive monkeys and kittens have receptive field properties very much like those of the adults. However, there are differences in their visual systems, such as in layer 4 of the striate cortex, where the projections from the LGN terminate. At birth, the cells in layer 4 are driven by both eyes, as projections from the LGN spread over a wide region of layer 4, whereas in the adult a layer 4 cell is driven by either eye but not by both. The adult pattern of ocular dominance columns in layer 4 is established over the first 6 weeks of life, when the LGN axons retract to establish separate, alternating zones in layer 4 which are supplied exclusively by one eye or the other.

Monocular or binocular deprivation

Rearing kittens with one eye sutured causes a series of changes throughout the visual system and drastically reduces the perceptual capabilities of the eye that had been sutured during early development (Wiesel & Hubel, 1963*a*, 1965). In the LGN, neurons connected to the deprived eye were reduced in size by 40% relative to the neurons connected to the other eye (Wiesel & Hubel, 1963*b*). Further, studies on the terminal fields of the reduced LGN cells in layer 4 showed that LGN axons connected to the deprived eye occupied less than 20% of the cortical area, and the other non-deprived eye had expanded its representation to cover more than 80% of the thalamic recipient zone (LeVay, Stryker & Shatz, 1978). Single-unit recording studies have shown that stimuli presented through the formerly deprived eye failed to influence the majority of cells in the striate visual cortex. The undeprived eye becomes the primary effective route for visual stimuli (see Figure 5.1).

Under conditions of dark rearing, the organization of the visual system and the selectivity of the cells initially continues to develop, despite the lack of visual stimuli (Buisseret & Imbert, 1976). When both eyes are closed in newborn monkeys for 17 days or longer, most cortical cells (such as the simple and complex cells) respond largely as normal to visual stimuli (Daw *et al.*, 1983). The organization of layer 4 seems to be normal and in other layers, most cortical cells are stimulated by both eyes. The major difference is that a large proportion of cells could not be driven binocularly, some

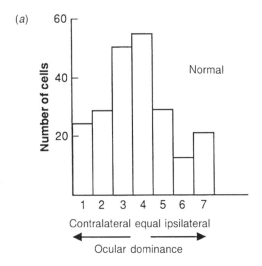

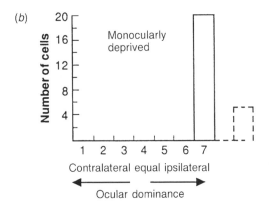

Fig. 5.1. (*a*) Ocular dominance histogram for 223 cells recorded from the visual cortex of adult cats. Cells in groups 1 and 7 of the histogram are driven by one eye only (ipsilateral or contralateral). All the other cells have inputs from both eyes. In Groups 2, 3, 4 and 5, the input from one eye is dominant. In Group 4 both eyes have a roughly equal influence. (*b*) Ocular dominance histogram of 25 cells recorded from visual cortex of a $2\frac{1}{2}$ month old kitten that was reared with its right eye occluded until the time of the experiment. The dashed bar on the right indicates that five cells did not respond to the stimulation of either eye. The solid bar indicates that all 20 cells that were responsive to stimulation responded only to the eye that was opened during rearing (redrawn from Wiesel & Hubel, 1963).

spontaneously driven cells could not be driven at all, while others were less tightly tuned to stimulus orientation.

Binocular deprivation in kittens leads to similar results, except that more cortical cells continue to be binocularly driven (Wiesel & Hubel, 1965). Longer visual deprivation (3 months or more) leads to a more marked effect. The visual cells become weakly responsive or totally unresponsive to visual stimuli, and the weakly responding cells lack orientation, direction and stereoselectivity (Cyander *et al.*, 1976; Imbert & Buisseret, 1976; Sherk & Stryker, 1976; Pettigrew, 1974). It seems that some of the results of monocular deprivation can be prevented or reduced by binocular deprivation. It may be that the two eyes are competing for representation in the cortex, and with one eye closed the contest becomes unequal.

What then, is the physiological basis for this ocular dominance shift associated with monocular deprivation? Such a shift can be prevented by modifying neuromodulator and neurotransmitter functions in the cortex (e.g. Shaw & Cyander, 1984; Bear & Singer, 1986; Reiter & Stryker, 1988), for example, infusion of glutamate into the cortex for a two-week period during monocular deprivation. Control recordings during the infusion period show that cortical neurons in general fail to respond well to visual stimuli from either eye during the infusion period. The lack of ocular dominance modification seems to be due to the reduced ability of the cortical cells to respond to the unbalanced LGN afferent input. Effective inputs representing the two eyes is greater than that of the deprived eye. It seems that, although changes associated with monocular deprivation have been found at the level of retinogeniculate terminals, lateral geniculate body somata, in LGN terminal and cortical cells' responses, the primary site of binocular competition is cortical, and other changes in the visual system are secondary to the primary cortical competition.

Competitive binocular interactions seem to occur within a sensitive period of development. Deprivation for only a day between the fourth and fifth week causes a large change in the pattern of ocular dominance (see Fig. 5.2) (Hubel & Wiesel, 1970; Olson & Freeman, 1980). If deprivation was started past about the eighth week, smaller effects were observed, until even long periods of deprivation at 4 months caused only a small effect (Fig. 5.3). Hubel and Wiesel concluded that the sensitive period for susceptibility to monocular deprivation begins in the fourth week and extended to about 4 months of age. Deprivation does not have to be long to cause large effects if it occurs during the sensitive period. Occluding one of the eyes of a 4-week-old kitten for a single day causes a large effect on the pattern of ocular dominance (Olson & Freeman, 1975). However, it seems

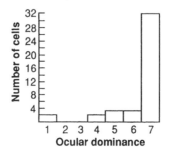

Fig. 5.2. Ocular dominance histogram of a kitten that had one eye occluded for 24 hours following four weeks of normal vision (redrawn from Olson & Freeman, 1975).

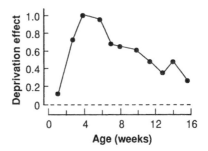

Fig. 5.3. Profile of the sensitive period for monocular deprivation in kittens. As can be seen, the most sensitive period is at 4–6 weeks, but monocular deprivation can cause substantial effects as long as 4 months after birth (redrawn from Olson & Freeman, 1980).

that the sensitive period is not fixed (Cyander & Mitchell, 1980; Cyander, 1983; Mower *et al.*, 1981). If cats are reared in the dark until long after the end of the chronologically defined sensitive period, and only then brought into the light for monocular deprivation, it can still produce marked effects on cortical ocular dominance. Dark reared cats seem to undergo a new sensitive period in the first few weeks after they are brought into the light.

Image misalignment and binocularity

Misalignment of the images in the two eyes can be accomplished either by cutting the eye muscles or by fitting the animal with a helmet that contains small optical prisms. Both procedures cause effects similar to those produced by monocular deprivation. Whereas 80% of cortical cells in normal cats are binocular, only 20% of the cells in cats with cut eye muscles respond to the stimulation of both eyes (Hubel & Wiesel, 1965). Similarly,

70% of cortical cells in monkeys are binocular, but less than 10% of cells are binocular in a monkey which has worn a prism-helmet for 60 days (Crawford & von Noorden, 1980). These neurological changes translate into striking behavioural effects. Prism-reared monkeys are unable to detect depth in random dot stereograms, suggesting that they have lost the ability to use binocular disparity to perceive depth (Crawford *et al.*, 1984).

This cortical malleability allows binocular neurons to adjust to the increased spacing between the eyes that occurs as animals grow. For a neuron to be able to signal binocular disparity, its receptive field in the left eye must be in exactly the right relation to its receptive field in the right eye. As an animal grows, the spacing between the eyes changes, so two receptive fields that would be lined up for closely spaced eyes would be thrown out of alignment for widely spaced eyes. Cortical malleability allows neurons to compensate for this slow but substantial change in eye spacing, thereby enabling the receptive fields to maintain their appropriate relationship to one another. Consistent with this hypothesis are the results of deprivation experiments from animals with frontal and lateral eyes. The degree of overlap is far greater in animals with front facing eyes, such as cats and monkeys, than in animals with lateral eyes, like rabbits or rats. The visual cortex of animals with lateral eyes is only slightly affected by deprivation. It has been suggested that the visual cortex of a lateral-eyed animal lacks malleability because its neurons do not need to be constantly recalibrated as the spacing between the eyes changes (Mitchell & Timney, 1984).

Selective rearing

To determine if visual experience plays an active role in shaping the development of visual neurons, vertical or horizontal stripes were used, either by placing cats in striped tubes (e.g. Blakemore & Cooper, 1970) or by fitting them with goggles that presented vertical stripes to one eye and horizontal stripes to the other (e.g. Hirsch & Spinelli, 1970, 1971).

Blakemore and Cooper kept kittens in the dark from birth to 2 weeks of age and then placed them in the tube for 5 hours every day. For the rest of the day they remained in the dark. The kittens sat on a plexiglass floor and the tube extended above and below them to ensure there were no visible corners or edges in their environment other than the stripes on sides of the tube. The kittens wore neck ruffs to prevent them altering the perceived orientation of the stripes by turning their heads. After 5 months, the selective rearing was stopped and the kittens remained in the dark except for brief periods when their vision was tested. The kittens displayed a number of defects in their visual behaviour. Their head movements were jerky when following moving objects, they tried to touch distant objects and were often unable to navigate around objects in their environment. Moreover, they seemed to be blind to stripes orthogonal to the orientation of the environment in which they were reared. Following these behavioural tests, Blakemore and Cooper recorded from cells in the visual cortex to determine the optimum stimulus orientation for different cells. Most of the cells of the horizontal stimuli-reared cats responded primarily to horizontal stimuli and none at all responded to vertical stimuli. The opposite is true of the vertical stimuli-reared cats. These results have been confirmed by subsequent experiments (Hirsch, 1972; Muir & Mitchell, 1975; Blasdel et al., 1977). The results of Hirsch and Spinelli's experiments using goggles showed the same pattern of effect. In single cell recording experiments they found few cells in the visual cortex whose preferred orientation deviated from the orientation of the environmental stimulus by more than 5–10 degrees.

A single hour of exposure in a striped tube can drastically alter the preferred orientation of cells in the visual cortex. Blakemore and Mitchell (1973) kept a kitten in the dark until 28 days of age, placed the kitten in a tube with vertical stripes for a single hour, and then kept the kitten in the dark until they recorded from its visual cortex at 42 days of age. As with kittens exposed to vertical stripes for much longer periods of time, most cells responded best to vertical or near-vertical orientations.

A number of different types of environment have been used to explore cortical plasticity further, such as moving white spots, random arrays of point sources of light and stripes moving in one particular direction (Van Sluyters & Blakemore, 1973; Pettigrew & Freeman, 1973; Tretter, Cyander & Singer, 1975; Cyander et al., 1975). In each case, the majority of cortical cells responded to the stimuli which were present in their environment, and very weakly to anything else. An interesting example of selective rearing is shown in cats reared under conditions of stroboscopic illumination, where continuous retinal movement is prevented. This results in a deficit in the selectivity of the cortical cells to direction of motion of visual stimuli (Cyander & Chernenko, 1976; Pasternak et al., 1985). The development of other cortical cell properties, such as orientation and stereoselectivity, remains unaffected under these conditions.

Several suggestions have been made as to the functional reasons for this sensitivity to environmental factors. The instructive acquisition hypothesis suggests that the environment instructs a neuron's development, so it acquires the ability to respond in a certain way. Thus, when a cat is raised in a vertical environment, cells that originally responded best to other than vertical orientations are 'instructed' to change into cells that respond best to verticals. Consistent with this hypothesis is the finding that, if an electrode is advanced obliquely through the cortex of a kitten that has been reared viewing only vertical contours, it encounters long sequences of vertically selective neurons (Rauschecher & Singer, 1981). In a normal cat, an oblique penetration would encounter neurons that respond to many orientations as the electrode passes through many orientation columns.

However, this theory has its limits. Although experience can alter a cell's orientation preference, a neuron cannot adopt a preference for an orientation to which it does not respond at all. For example, it is possible to create a cell with a vertical stimulus preference if it originally responded to a vertical stimulus to some small degree, but if it never responded to a vertical stimulus in the first place, it cannot be converted to responding to a vertical stimulus. An alternative theory is the selective degeneration hypothesis, which postulates that neurons that are not stimulated by a particular environment lose their ability to respond, leaving only those neurons which are stimulated by the rearing environment. Consistent with this explanation is the finding that cats reared with horizontal/vertical goggles have many cells that respond abnormally, or become completely unresponsive (Stryker *et al.*, 1978). These unresponsive cells were probably not stimulated by the kitten's selective environment.

Active environmental interactions

It is not enough for an animal merely to observe its environment passively. For an animal to actually develop the sensory–motor connections in order to use visual information, the animal must interact with its environment. For example, this has been tested in dark-reared kittens given their first visual experience in the 'kitten carousel' (Held & Hein, 1963). As an 'active' kitten walks, it transports a 'passive' kitten in a gondola. In this way both kittens receive the same visual stimulation, but this stimulation is self-produced motion for only the active kitten. The active kitten appears to develop normal sensory–motor co-ordination. It blinks at approaching objects and puts out its paws to ward off a collision when carried towards a surface. The passive kitten shows neither of these reactions, and is considered to possess poor sensory–motor co-ordination.

In humans, the visual system seems to remain quite plastic even in the adult. If an adult wears prism goggles which shift perception to the left or right, the individual will initially show poor sensory–motor co-ordination. The individual's perception of the environment has been artificially shifted with respect to its actual position. However, the individual eventually adapts to the displaced perception and can function normally. This again requires the experience of self-produced movement. If a person wearing goggles walks along a path, while a second person also wearing goggles is pushed along the path in a wheelchair, whereas both experience the similar visual stimulation, only the active person will adapt to the goggles (Held, 1965).

The visual system and the environment

The complexity of the visual environment in which an animal lives, both during development and once mature, has a profound effect on its visual system and associated areas. All the evidence shows that an impoverished visual environment has a detrimental effect on the normal development of the visual system, particularly in a sensitive period just after birth. If these impairments are not corrected rapidly, they seem to be largely irreversible. Moreover, the adult visual system is constantly updating its responses to accommodate new objects and visual scenes as it encounters them. This is represented in changes in response properties at the single cell level (Rolls, Tovée & Ramachandran, 1993), and it has been suggested that an impoverished visual environment may reduce this flexibility. The laboratory environment, which is designed with hygiene and ease of cleaning as a priority, is often very bare and sparse in terms of visual stimulation. This may facilitate the maintenance of an animal's gross health, but it may have a detrimental effect on its visual nervous system. Moreover, it is not sufficient for an animal merely to observe an environment; for the normal sensory–motor connections to be made, an animal must interact with its surroundings. With space at a premium in many laboratories, cage space is often the minimum required by law. This approach may be short-sighted. The provision of an exercise area to which the animal has access at least part of the day will improve not only its quality of life but also the state of its nervous system.

A higher-order processing system: the hippocampus

The neurophysiological effects of environmental enrichment, or impoverishment, on areas other than sensory areas has attracted only sporadic attention. This is not terribly surprising, given the nature of the research required to investigate such potential neurophysiological effects; those interested, for example, in issues of animal welfare are unlikely to find this field of research particularly attractive. There is, however, a body of literature on the nature of such effects on laboratory animals, particularly rats.

Within a week of housing rats in an enriched environment (variety of toys and several conspecifics) gross morphological changes can be observed in the brain (Rosenzweig & Bennett, 1978). Over longer periods total brain size can increase by about 5%, with the largest changes occurring in the cortex (8–9% in the occipital region). This increase in brain size does not seem to be due to a greater number of neurons, indeed neuron density is lower in rats housed in enriched environments than in those housed in an impoverished environment (Diamond, Krech & Rosenzweig, 1964; Beaulieu & Colonnier, 1985). Rather, the brain size variation appears to be due to variation in the number of glial cells (Diamond *et al.*, 1966; Szeligo, 1977; Szeligo & Leblond, 1977). It has been suggested that this is due to the role the glial cells play in metabolic support for neuronal activity. Proliferation of dendritic branching also contributes to the growth in brain size (Holloway, 1966; Volkmar & Greenhough, 1972; Globus *et al.*, 1973; Uylings *et al.*, 1978).

In general, large and predictable changes in whole brain structure and function do occur in response to environmental manipulation. These effects are most noticeable in the brains of animals provided with an enriched environment. Several explanations for these neurophysiological changes have been advanced: (a) variation in the animals' maturation rate; (b) variation in stress levels; (c) hormonal variation; (d) social stimulation; (e) visual stimulation; (f) learning and memory. All but the last of these have been shown to be unsatisfactory interpretations, for example, housing rats in groups without toys is not sufficient to produce the observed changes, and animals housed in impoverished conditions for over 500 days do not eventually show brain size increases (Cummins *et al.*, 1973; Rosenzweig & Bennett, 1978). Visual stimulation, discussed in the previous section, is also not sufficient to explain these gross brain changes: no changes were observed in animals permitted simply to view an enriched environment (Ferchmin, Bennett & Rosenzweig, 1975). Possibly the explanation with the greatest intuitive appeal is that learning and memory account for the changes in neural structures.

The hippocampus is the area of the brain currently attracting more interest than any other in relation to learning and memory. There are several reasons for the intensity of interest: first, and possibly foremost, is the major role the hippocampus plays in human memory. Damage to the human hippocampus results in serious anterograde amnesia. It appears as if such damage disrupts consolidation of short-term memories into long-term storage. A second reason for the interest in the hippocampus stems from the finding of long-term potentiation (LTP). Bliss and co-workers (1973) found that brief tetanic stimulation of afferents from entorhinal cortex to the hippocampus in a rabbit preparation followed by a second burst of stimulation resulted in an increase in the magnitude of response observed at the post-synaptic terminal (Bliss & Lmo, 1973; Bliss & Gardner-Medwin, 1973). This increase in response magnitude was stable and lasted for several weeks. These findings were extremely important as long-term potentiation seemed to be the long-awaited candidate for a mechanism underlying vertebrate learning.

The question arises as to whether there are changes observed in hippocampal structure and/or function in response to environmental conditions? Although Walsh *et al.* (1969) observed a thickening of the hippocampus of rats given an enriched environment, subsequent studies have found little evidence of major morphological changes (Rosenzweig & Bennett, 1978; Jones & Smith, 1980). This is, perhaps, surprising given the marked changes seen in the cortex in response to the same kind of environmental manipulation (e.g. Rosenzweig, Bennett & Diamond, 1972*a,b*). There may not be much evidence for major volumetric changes in the hippocampus but there are, however, convincing data for changes in both the number of neurons (granule cells) and dendritic branching in the dentate gyrus (one of two interlocking parts of the mammalian hippocampal formation): both have been shown to increase in rat hippocampus in response to an enriched environment (Susser & Wallace, 1982; Fiala, Joyce & Greenhough, 1978). In addition, Altschuler (1979) found increases in synaptic density. It is not known whether these features show a decrease with exposure to an impoverished environment. Changes in the electrophysiological properties of the rat dentate gyrus also occur in response to an enriched environment. Sharp, McNaughton and Barnes (1985) housed rats singly in small cages for 10 days, during which time they recorded responses from granule cells to single stimulus pulses. Some of the rats were then transferred into individual rooms with boxes, ramps and other surfaces for the rats to move about on. Following the transfer to the spatially enriched environment, there was an increase in the magnitude of the field excitatory post-synaptic potential (EPSP) and in the population spike amplitude. These data were exciting in the light of a proposed role for the hippocampus in the processing of spatial information put forward by O'Keefe and Nadel (1978). They suggested that the hippocampus was the site controlling allocentric spatial learning, in particular the learning and computation of novel routes within familiar areas. Formerly, the finding of place cells, cells within the hippocampus which fire according to an animal's location, had been the strongest evidence in favour of this hypothesis (O'Keefe & Dostrovsky, 1971). This discovery of long-term enhancement (LTE, a phenomenon similar to LTP) was an exciting result, as it suggested not only a relationship between the processing of spatial information and the hippocampus but also a mechanism for the translation from the acquisition of spatial information into spatial memory.

Novelty or space?

It is not clear from Sharp *et al.*'s (1985) data whether the observed changes occur in response to the novelty of the enriched environment or to the alterations in the spatial component of the new environment. Most experiments manipulating environmental richness have concentrated on the provision of toys and conspecifics with the former being changed on a regular basis. Somewhat surprisingly, Sharp *et al.* (1985; Sharp, McNaughton and Barnes, 1987) appear to have been the first to examine the neurophysiological effects of modifying spatial features, such as cage size, in spite of the likelihood that spatial considerations play a large role in the design and provision of ideal housing conditions. The suggestion that the size of an area that an animal regularly moves around (such as a cage or a territory) has an effect on the morphology of the hippocampus comes from data on hippocampus size of domesticated species and ancestral wild races. Hippocampus size, and that of the whole brain, is smaller in domesticated individuals than in their wild relatives (Kruska, 1988). If the hippocampus is processing spatial information, and if domestic animals are spatially restricted in comparison with their wild counterparts, the variation in hippocampal volume could be explained by a change in the strength of selection on the hippocampus. A decreased selective pressure on hippocampus size should lead to a diminishing of its processing capacity and then perhaps to a decrease in hippocampal volume. Support for this proposal comes from a study on hippocampus size and range size in two species of vole. In meadow voles, *Microtus pennsylvanicus*, the male has a substantially larger home range size than the female whereas in pine voles, *M. pinetorum*, males and females have similarly sized ranges. Interestingly, although perhaps coincidentally, the male meadow vole has a much larger hippocampus than conspecific females, while male pine voles do not differ from females in hippocampal volume (Jacobs *et al.*, 1990). On the other hand, there is no evidence that hippocampus size, or whole brain size, increases when domesticated animals are returned to the wild, even after several generations (Kruska, 1987). It is not clear why there have not been brain changes in these previously domesticated animals, although it may be simply that there has not been sufficient time for changes, large enough to be measured, to have occurred.

The question remains as to whether the provision or experience of a larger area *per se* is sufficient to bring about changes in the hippocampus? Alternatively, whether the environmental manipulations which involved the addition of features also enhanced spatial complexity? No experiment, to our knowledge, has yet directly distinguished between the effects contributed by these two factors. However, an experiment conducted by Clayton and Krebs (1994) may shed some light on this problem. Clayton and Krebs were interested in the effect of food storing experience on hippocampus size in the juvenile marsh tits (*Parus palustris*), a food-storing passerine species. Previously, adult food storers had been shown to have larger hippocampal volumes, relative to the rest of the telencephalon, than young pre-storing individuals (Healy & Krebs, 1993; Healy, Clayton & Krebs, 1994). This difference was not observed in two non-storing species. Clayton and Krebs hand-raised juvenile marsh tits and these birds were allowed either to store and retrieve in an experimental room or simply to feed and fly about in this room. The birds with the food storing experience had significantly larger hippocampal volumes than did the control birds. The crucial difference appears not to be a mechanical one (e.g. amount of flying experience was equivalent between the groups) but rather the learning of, and memory for, locations specifically associated with food. In order to retrieve food accurately as these birds did, they had to remember the spatial relationship of the available or salient visual landmarks relative to (or associated with) the storage sites and of storage sites to each other. The birds that simply exercised in the same room need pay little attention to these landmarks and thus may not have consigned them to memory in any great detail. The results of this experiment may suggest that other animals provided with the appropriate motives to learn new locations or routes, usually food, may also show changes in the structure or functioning capacity of the hippocampus. A study on brood-parasitic brown-headed cowbirds (*Molothrus ater*) lends credence to this hypothesis (Sherry *et al.*, 1993). Females of this species search for host nests in which to lay their eggs and return within several days to lay an egg.

They may lay eggs in as many as 40 host nests, presumably using memory for relocation of these nests. It is significant that male brown-headed cowbirds do not aid the females in locating nests and they have significantly smaller hippocampal volume than do females.

This apparent relationship between the hippocampus and the spatial complexity of an animal's environment may go some way towards explaining why so few studies thus far have found an effect of enrichment on hippocampal structure and function. As noted earlier, most of the enrichment or impoverishment studies conducted so far have involved the addition or removal of conspecifics or toys. Whilst the novelty of, and interactions with, these additions will require the acquisition of new information, they are not of a spatial nature. Providing features which are spatially fixed, like the tunnels and ramps in Sharp *et al.* (1985, 1987), does seem to bring about changes in the hippocampus. As with the development of the visual system, the active involvement of the animal with its environment seems to be necessary in bringing about hippocampal changes. The role of active involvement is supported still further by the data from place cell firing: place cells, situated in the hippocampus, fire depending on the animal's location (O'Keefe & Dostrovsky, 1971). It is not known what the effects might be of removing such features, as might approximate the situation of a wild-caught animal, nor the effects of never having provided them, as in the case of most laboratory-reared animals. Perhaps, in the case of wild-caught animals, this removal would result in a major reduction in place cell firing and, in turn, the gradual loss of synaptic connections both within, and leading to and from, the hippocampus. Laboratory-reared animals that have never experienced a complex spatial environment, perhaps, will never form such connections. Whether the ability to form such connections is also lost is not known.

Effects of ageing

Whilst investigating possible neurophysiological changes and the role of the environment, one needs to be aware of at least two opposing effects of increasing age. The first of these is that many animal species have substantial post-natal brain growth. This is especially true of those species which produce altricial young. Their young are produced after a relatively short gestation and have a smaller brain size at birth than precocial young, although adult brain size is not correlated with development strategy (humans being the exception) (Harvey, Martin & Clutton-Brock, 1987). Large-scale post-natal brain size changes then will be much more apparent in species with altricial young. Whilst there are few data showing whether environmental input can modify this post-natal growth curve, a recent study looked at the effects of ageing and captivity on hippocampus size in garden warblers (*Sylvia borin*), a migratory passerine species (Healy, Gwinner & Krebs, 1996). Birds were caught at 3 months old and kept in captivity for 12 months. These 15-month-old birds had significantly larger relative hippocampal volumes than 3-month-old birds, which had never been held in captivity. Thus, in spite of the restrictive spatial conditions, there was a substantial age-related increase in hippocampal volume.

Another, perhaps more obvious, consistent effect of ageing is the decline in performance on spatial tasks in animals and humans of advanced age (e.g. Evans *et al.*, 1984; Barnes, Nadel & Honig, 1980). Concomitant changes are also seen in the electrophysiological properties of the ageing rat hippocampus, but it is not clear if there are also major structural changes. The acquisition rate of spatial memory problems by ageing rats is slower and the forgetting rate is faster than in young rats. The increase in LTE and its subsequent decline shows a similar pattern (Barnes & McNaughton, 1985). It is not yet known whether prior experience has any impact on the rate of change of these processes.

In summary, consideration of the current data investigating the impact of environmental complexity on hippocampal structure and function raises many more questions than can be answered adequately at present. This is perhaps due to the nature of most of the experiments conducted, to date, on environmental enrichment or impoverishment (and there are few of the latter). Variation in the spatial aspects of an animal's environment has been rarely examined explicitly, and even less often has there been consideration of the effect this might have on the region of brain implicated in the processing of spatial information. Evidence gathered thus far, however, not only shows a correlation between variation in hippocampal volume and spatial complexity but also experience-dependent volumetric and electrophysiological changes. These data are exciting because they provide the first signs that substantial neurophysiological changes can occur in adult brains as a result of experience in areas other than those processing sensory information.

Summary

The question we originally asked was: 'What is the neurophysiological impact of variation in environmental input?' Specifically, can the structure and function of brain regions be modified by the provision or removal of different stimuli or by requiring an animal to perform particular tasks? The answer to the second question is a resounding 'yes': brain structure and function can be modified, in some cases drastically, as a result of, for example, varying the visual stimuli the animal is exposed to, or by the provision of housing features such as tunnels and ramps. The impact of restricting environmental input has been explored most fully with the visual system and radical, irrevocable changes may result, the extent of which is largely dependent on the animal's developmental stage. The most striking consequences are observed in the brains of juveniles and these are often restricted to a sensitive period. Such sensitive periods are limited to specific lengths of time, and environmental manipulations occurring outside of the sensitive period may bring about little or no change.

The effects of spatial restriction and reduced spatial complexity, however, as often occurs by the caging of animals, are largely unknown. The few tests which have been performed suggest that increasing spatial complexity and/or decreasing spatial restriction does result in both structural and functional changes in higher processing areas like the hippocampus. It is probable that the majority of these changes are restricted to the developing brain, as in the sensory systems, although there is little evidence for a sensitive period. Whether such changes are reversible is an open question. If the timing of the post-natal development of higher processing areas is like that of the sensory systems, then it is the environment that the young animal is exposed that is of major importance. The environmental impact on the adult brain, on the other hand, is likely to be of little consequence in neurophysiological terms.

Finally, we would like to draw attention to a point raised by Renner and Rosenzweig (1987). They called attention to the standard laboratory animal's housing conditions which increasingly stringent regulations now usually demand animals to be housed singly, in easy-to-clean cages with a constant excess supply of food and water. As we have seen that conditions such as these seem to lead to a reduction in both the size and functioning capabilities of different brain regions. It seems somewhat ironic that much of what we know about, for example, learning and memory in mammals, comes from animals (particularly rats and monkeys) housed in just such environments.

References

Altschuler, R. A. (1979). The effects of increased experience and training of synaptic density in Area CA3 of the rat hippocampus. *Dissertation Abstract*

International, **39**,(9B), 4140–1.

Barnes, C. A. & McNaughton, B. L. (1985). An age comparison of the rates of acquisition and forgetting spatial information in relation to long-term enhancement of hippocampal synapses. *Behavioural Neuroscience*, **99**, 1040–8.

Barnes, C. A. , Nadel, L. & Honig, W. K. (1980). Spatial memory deficit in senescent rats. *Canadian Journal of Psychology*, **34**, 29–39.

Bear, M. & Singer, W. (1986). Modulation of visual cortex plasticity by acetylcholine and noradrenaline. *Nature*, **320**, 172–6.

Beaulieu, C. & Colonnier, M. (1985). The effects of impoverished and enriched environments on the number and size of boutons containing flat vesicles in the visual cortex of cat. *Society of Neuroscience Abstracts*, **12**, 128.

Bedi, K. S. & Bhide, P. G. (1988). Effects of environmental diversity on brain morphology. *Early Human Development*, **17**, 107–43.

Blakemore, C. & Cooper, G. F. (1970). Development of the brain depends on the visual environment. *Nature*, **228**, 477–8.

Blakemore, C. & Mitchell, E. D. (1973). Environmental modification of the visual cortex and the neural basis of learning and memory. *Nature*, **241**, 467–8.

Blasdel, G. G. , Mitchell, D. E. , Muir, D. W. & Pettigrew, J. D. (1977). A combined physiological and behavioural study of the effect of early visual experience with contours of a single orientation. *Journal of Physiology*, **265**, 615–36.

Bliss, T. V. P. & Gardner-Medwin, A. R. (1973). Long-lasting potentiation of synaptic transmission in the dentate area of the unanaesthetized rabbit following stimulation of the perforant path. *Journal of Physiology*, **232**, 357–74.

Bliss, T. V. P. & L·mo, T. (1973). Long-lasting potentiation of synaptic transmission in the dentate area of the anaesthetized rabbit following stimulation of the perforant path. *Journal of Physiology*, **232**, 331–56.

Buisseret, D. & Imbert, M. (1976). Visual cortical cells: their development properties in normal and dark reared kittens. *Journal of Physiology*, **255**, 511–25.

Clayton, N. S. & Krebs, J. R. (1994). Hippocampal growth and attrition in birds affected by experience. *Proceedings of the National Academy of Sciences, USA*, **16**, 7410–14.

Crawford, M. L. J. & von Noorden, G. K. (1980). Optically induced concomitant strabismus in monkey. *Investigations in Ophthalmology and Vision Science*, **19**, 1105–9.

Crawford, M. L. J. , Smith, E. L. , Harwerth, R. S. & von Noorden, G. K. (1984). Stereoblind monkeys have few binocular neurons. *Investigations in Ophthalmology and Vision Science*, **25**, 779–81.

Cummins, R. A. , Walsh, R. N. , Budtz-Olsen, O. E. , Konstantinos, T. & Horsfall, C. R. (1973). Environmentally-induced changes in the brains of elderly rats. *Nature*, **243**, 516–18.

Cyander, M. (1983). Prolonged sensitivity to monocular deprivation in dark reared cats: effects of age and visual exposure. *Journal of Neurophysiology*, **43**, 1026–40.

Cyander, M. & Chernenko, G. (1976). Abolition of directional selectivity in the visual cortex of the cat. *Science*, **193**, 504–5.

Cyander, M. & Mitchell, D. E. (1980). Prolonged sensitivity to monocular deprivation in dark reared cats. *Journal of Neurophysiology*, **43**, 1026–40.

Cyander, M. , Berman, N. & Hein, A. (1975). Cats raised in a one directional world: effects on receptive fields in visual cortex and superior colliculus. *Experimental Brain Research*, **22**, 267–80.

Cyander, M. , Berman, N. & Hein, A. (1976). Recovery of function in cat visual cortex following prolonged deprivation. *Experimental Brain Research*, **25**, 139–56.

Daw, N. W. , Rader, R. K. , Robertson, T. W. & Ariel, M. (1983). Effects of 6-hydroxy-dopamine on visual deprivation in the kitten cortex. *Journal of Neuroscience*, **3**, 907–14.

DeValois, R. L. , Albrecht, D. G. & Thorell, L. (1978). Cortical cells: bar detectors or spatial frequency filters? In *Frontiers In Visual Science*, ed. S. J. Coll, & Smith. E. L. , Berlin: Springer-Verlag.

Diamond, M. C. , Krech, D. & Rosenzweig, M. R. (1964). The effects of an enriched environment of the histology of the rat cerebral cortex. *Journal of Comparative Neurology*, **123**, 111–20.

Diamond, M. C. , Law, F. , Rhodes, H. , Lindner, B. , Rosenzweig, M. R. , Krech, D. & Bennett, E. L. (1966). Increases in cortical depth and glia numbers in rats subjected to enriched environments. *Journal of Comparative Neurology*, **128**, 117–26.

Evans, G. W. , Brennan, P. L. , Skorpanich, M. A. & Held, D. (1984). Cognitive mapping and elderly adults: verbal and location memory for urban landmarks. *Journal of Gerontology*, **39**, 452–7.

Ferchmin, P. A. , Bennett, E. L. & Rosenzweig, M. R. (1975). Direct contact with enriched environment is required to alter cerebral weights in rats. *Journal of Comparative Physiology and Psychology*, **88**, 360–7.

Fiala, B. , Snow, F. M. & Greenhough, W. T. (1977). Impoverished rats weigh more than enriched rats because they eat more. *Developments in Psychobiology*, **10**, 537–41.

Fiala, B. A. , Joyce, J. N. & Greenhough, W. T. (1978). Environmental complexity modulates growth of granule cell dendrites in developing but not adult hippocampus of rats. *Experimental Neurology*, **59**, 372–83.

Globus, A. , Rosenzweig, M. R. , Bennett, E. L. & Diamond, M. C. (1973). Effects of differential environments on dendritic spine counts. *Journal of Comparative Physiology and Psychology*, **82**, 175–181.

Harvey, P. H. , Martin, R. D. & Clutton-Brock, T. H. (1987). Life histories in comparative perspective. In *Primate Societies*, ed. N. N. Smuts, D. L. Cheney, R. M. Seyfarth, R. W. Wrangham & T. T. Struhsaker, pp. 181–96. Chicago: Chicago University Press.

Healy, S. D. & Krebs, J. R. (1993). Development of hippocampal specialisation in a food-storing bird. *Behavioural Brain Research*, **53**, 127–31.

Healy, S. D. , Clayton, N. S. & Krebs, J. R. (1994). Development of hippocampal specialisation in two species of tit (*Parus* spp.). *Behavioural Brain Research*, **61**, 23–8.

Healy, S. D. , Gwinner, E. & Krebs, J. R. (1996). Hippocampal volume in migrating and non-migrating warblers: effects of age and experience. *Behavioural Brain Research*, **81**, 61–8.

Held, R. (1965). Plasticity in sensory–motor systems. *Scientific American*, **213**, 84–94.

Held, R. & Hein, A. (1963). Movement produced stimulation in the development of visually guided behaviour. *Journal of Comparative Physiology and*

Psychology, **56**, 872–6.

Hirsch, H. V. B. (1972). Visual perception in cats after environmental surgery. *Experimental Brain Research*, **15**, 405–23.

Hirsch, H. V. B. & Spinelli, D. N. (1970). Visual experience modifies distribution of horizontally and vertically orientated receptive fields in cats. *Science*, **168**, 869–71.

Hirsch, H. V. B. & Spinelli, D. N. (1971). Modification of the distribution of receptive field orientations in cat by selective visual exposure during development. *Experimental Brain Research*, **12**, 509–27.

Holloway, R. L. (1966). Dendritic branching in the rat visual cortex. Effects of extra environmental complexity and training. *Brain Research*, **2**, 393.

Hubel, D. H. & Wiesel, T. N. (1965). Binocular interaction in straite cortex of kittens reared with artificial squint. *Journal of Neurophysiology*, **28**, 1041–59.

Hubel, D. H. & Wiesel, T. N. (1970). The period of susceptibility to the physiological effects of unilateral lid closure in kittens. *Journal of Physiology*, **206**, 419–36.

Hubel, D. H. & Wiesel, T. N. (1971). Aberrant visual projections in the Siamese cat. *Journal of Physiology*, **218**, 33–62.

Imbert, M. & Buisseret, P. (1976). Receptive field characteristics and plastic properties of visual cortical cells of kittens reared with or without visual experience. *Experimental Brain Research*, **22**, 25–36.

Jacobs, L. F. , Gaulin, S. J. C. , Sherry, D. F. & Hoffman, G. E. (1990). Evolution of spatial cognition: sex-specific patterns of spatial behaviour predict hippocampal size. *Proceedings of the National Academy of Sciences, USA*, **87**, 6349–52.

Jones, D. G. & Smith, B. J. (1980). The hippocampus and its response to differential environments. *Progress in Neurobiology*, **15**, 19–69.

Kruska, D. (1987). How fast can total brain size change in mammals? *Journal für Hirnforschung*, **28**, 59–70.

Kruska, D. (1988). Mammalian domestication and its effect on brain structure and behavior. In *Intellegence and Evolutionary Biology*, ed. H. J. Jerison & I. Jerison, pp. 211–50. Berlin: Springer.

LeVay, S. , Stryker, M. P. & Shatz, C. J. (1978). Ocular dominance columns and their development of layer IV of the cat's visual cortex: a quantitative study. *Journal of Comparative Neurology*, **179**, 223–44.

Mitchell, D. E. & Timney, B. (1984). Postnatal development of function in the mammalian visual system. In *The Handbook of Physiology, The Nervous System III*, ed. M. Brookhart, & V. B. Mountcastle. Bethesda: American Physiological Society.

Mower, G. D. , Berry, D. , Birchfield, J. L. & Duffy, F. H. (1981). Comparison of the effects of dark rearing and binocular suture on development and plasticity of cat visual cortex. *Brain Research*, **220**, 255–67.

Muir, D. W. & Mitchell, D. E. (1975). Behavioural defects in cats following early selected visual exposure to contours of a single orientation. *Brain Research*, **85**, 459–77.

O'Keefe, J. & Dostrovsky, J. (1971). The hippocampus as a spatial map: preliminary evidence from unit activity in the freely moving rat. *Brain Research*, **34**, 171–5.

O'Keefe, J. & Nadel, L. (1978). *The Hippocampus as a Cognitive Map*. Oxford: Oxford University Press.

Olson, C. R. & Freeman, R. D. (1975). Progressive changes in kitten striate cortex during monocular vision. *Journal of Neurophysiology*, **38**, 26–32.

Olson, C. R. & Freeman, R. D. (1980). Profile of the sensitive period for monocular deprivation in kittens. *Experimental Brain Research*, **39**, 17–21.

Pasternak, T. , Schumer, R. , Gizzi, M. S. & Movshon, J. A. (1985). Abolition of cortical direction selectivity affects visual behaviour in cats. *Experimental Brain Research*, **61**, 214–17.

Pettigrew, J. D. (1974). The effect of visual experience on the development of stimulus specificity by kitten cortical neurons. *Journal of Physiology*, **237**, 49–74.

Pettigrew, J. D. & Freeman, R. D. (1973). Visual experience without lines: effects on developing cortical neurons. *Science*, **182**, 599–601.

Rauschecker, J. P. & Singer, W. (1981). The effects of early visual experience on the cats' visual cortex and their possible explanation by Hebb synapses. *Journal of Physiology*, **310**, 215–39.

Reiter, H. O. & Stryker, M. P. (1988). Neural plasticity without postsynaptic action potentials: less active inputs become dominant when kitten visual cortex cells are pharmacologically inhibited. *Proceedings of the National Academy of Sciences, USA*, **85**, 3623–7.

Renner, M. J. & Rosenzweig, M. R. (1987). *Enriched and Impoverished Environments. Effects on Brain and Behavior*. New York: Springer-Verlag.

Rolls, E. T. , Tovée, M. J. & Ramachandran, V. S. (1993). Visual learning reflected in the responses of neurons in the temporal visual cortex of the macaque. *Society for Neuroscience Abstracts*, **19**, 27.

Rosenzweig, M. R. & Bennett, E. L. (1978). Experiential influences on brain anatomy and brain chemistry in rodents. In *Studies on the Development of Behavior and the Nervous System*, ed. G. Gottlieb, pp. 289–327. New York: Academic Press.

Rosenzweig, M. R. , Bennett, E. L. & Diamond, M. C. (1972*a*). Cerebral effects of differential experience in hypophysectomized rats. *Journal of Comparative Physiology and Psychology*, **79**, 56–66.

Rosenzweig, M. R. , Bennett, E. L. & Diamond, M. C. (1972*b*). Brain changes in response to experience. *Scientific American*, **226**, 22–29.

Sharp, P. E. , McNaughton, B. L. & Barnes, C. A. (1985). Enhancement of hippocampal field potentials in rats exposed to a novel, complex environment. *Brain Research*, **339**, 361–5.

Sharp, P. E. , McNaughton, B. L. & Barnes, C. A. (1987). Effects of aging on environmental modulation of hippocampal evoked responses. *Behavioural Neuroscience*, **101**, 170–8.

Shaw, C. & Cyander, M. (1984). Disruption of cortical activity prevents alterations of ocular dominance in monocularly deprived kittens. *Nature*, **308**, 731–4.

Sherk, H. & Stryker, M. P. (1976). Quantitative study of cortical orientation selectivity in visually inexperienced kitten. *Journal of Neurophysiology*, **39**, 63–70.

Sherry, D. F. , Forbes, M. R. L. , Khurgel, M. & Ivy, G. O. (1993). Females have a larger hippocampus than males in the brood-parasitic brown-headed cowbird. *Proceedings of the National Academy of Sciences, USA*, **90**, 7839–43.

Stryker, M. P. , Sherk, H. , Leventhal, A. G. & Hirsch, H. B. (1978). Physiological consequences for the cat's visual cortex of effectively restricting early visual experience with oriented contours. *Journal of*

Neurophysiology, **41**, 869–909.

Susser, E. R. & Wallace, R. B. (1982). The effects of environmental complexity on the hippocampal formation of the adult rat. *Acta Neurobiologica Experimentalis*, **42**, 203–7.

Szeligo, F. (1977). Quantitative differences in oligodendrocytes and myelinated axons in the brains of rats raised in enriched, control, and impoverished environments. *Anatomical Record*, **187**, 726–7.

Szeligo, F. & Leblond, C. P. (1977). Response of the three main types of glial cells of cortex and corpus callosum in rats handled during suckling or exposed to enriched, control, or impoverished environments following weaning. *Journal of Comparative Neurobiology*, **172**, 247–64.

Tovée, M. J. (1994). The molecular genetics and evolution of primate colour vision. *Trends in Neurosciences*, **17**, 30–7.

Tretter, E. , Cyander, M. & Singer, W. (1975). Modification of direction selectivity of neurons in the visual cortex of kittens. *Brain Research*, **84**, 143–9.

Uylings, H. , Kuypers, K. , Diamond, M. & Veltman, W. (1978). Effects of differential environments on plasticity of dendrites of cortical pyramidal neurons in adult rats. *Experimental Neurology*, **62**, 658–77.

Van Sluyters, R. C. & Blakemore, C. (1973). Experimental creation of unusual properties in visual cortex of kittens. *Natrure*, **246**, 505–8.

Volkmar, F. R. & Greenhough, W. T. (1972). Rearing complexity affects branching of dendrites in the visual cortex of the rat. *Science*, **176**, 1447–9.

Walsh, R. N. , Budtz-Olsen, O. E. , Penny, J. E. & Cummins, R. A. (1969). The effects of environmental complexity on the histology of the rat hippocampus. *Journal of Comparative Neurology*, **137**, 361–6.

Wiesel, T. N. & Hubel, D. H. (1963a). Single-cell responses in striate cortex of kittens deprived of vision in one eye. *Journal of Neurophysiology*, **26**, 1003–17.

Wiesel, T. N. & Hubel, D. H. (1963b). The effects of visual deprivation on the morphology and physiology of cell's lateral geniculate body. *Journal of Neurophysiology*, **26**, 978–93.

Wiesel, T. N. & Hubel, D. H. (1965). Comparisons of the effects of unilateral and bilateral eye closure on single unit responses in kittens. *Journal of Neurophysiology*, **28**, 1029–40.

6

The behavioural requirements of farm animals for psychological well-being and survival

ROBERT J. YOUNG

Introduction

It was once thought that all a captive animal required was food, water and a mate. Scientists researching into the welfare of animals realize that their needs extend far beyond food, water and a mate (e.g. Hughes & Duncan, 1988; Dawkins, 1990; Poole, 1992). However, under current United Kingdom legislation, many farm animals are often deprived to varying degrees of these basic resources. This may not only affect animals' psychological well-being, but also their perception of their chances of survival.

This chapter will address the following issues that concern the abundance of areas in which the welfare of animals might be adversely affected:

(a) *Are there behavioural needs?*
(b) *Are the effects of depriving some behavioural patterns worse than others?*
(c) *How do we go about countering the effects of depriving animals the opportunities to perform their full behavioural repertoire?*

The rest of this chapter will address these questions.

The behavioural repertoires of animals have evolved to maximize the individual's fitness, which may be equated to the number of copies of its genes in the population (Krebs & Davies, 1993). An individual behavioural pattern may be divided into two distinct phases: appetitive and consummatory. For example, foraging behaviour consists of looking for food (appetitive behaviour) and eating the found food (consummatory behaviour). Since the consumption of food helps to restore physiological homeostasis with the animal's body, it is not surprising that such (consummatory) behaviour is positively reinforcing. However, some scientists interested in the welfare of animals have suggested that the appetitive phase (i.e. the looking for food) is also reinforcing (Herrnstein, 1977; Hughes & Duncan, 1988). This is based on Glickman and Schiff (1967, p. 81) *A Biological*

Theory of Reinforcement, which suggests: '... that reinforcement evolved as a mechanism to insure [*sic*] species-typical responses to appropriate stimuli'.

It has been suggested that the performance of appetitive behaviour may be as important as the succeeding consummatory behaviour (Hughes & Duncan, 1988; Dawkins, 1990). The idea that animals may need to perform some behavioural patterns for psychological well-being is often referred to as the animal having 'behavioural or ethological needs' (Hughes & Duncan, 1988). This idea is embodied in the UK Animal Welfare Codes, which state that animals have a 'need' to express most normal patterns of behaviour (MAFF, 1983).

Behavioural and physiological restrictions placed on farm animals

Food restriction is widely practised for the breeding stock of pigs (*Sus scrofa*) and broiler hens, *Gallus gallus domesticus* (Lawrence, Appleby & MacLeod, 1988; Savory, Maros & Rutter, 1993). Broiler hens are usually water restricted during the rearing period (Fraser & Broom, 1990). All farm animals when being transported or in lairage for slaughter, may be both food and water restricted for up to 24 hours (Fraser & Broom, 1990). Only the animals chosen by the farmer or the breeding company receive a mate. Most species of intensively farmed animals have little or no opportunity to perform foraging behaviour. Many intensive husbandry systems even deprive animals of the opportunity to adequately exercise their bodies. All of these practices are justified on economic grounds, for example, breeding female pigs (sows) are food restricted to save on feeding costs and to increase litter sizes (Whittemore, 1987). Further examples of behavioural and physical restrictions placed on the four main species of farm animals (pigs, hens, cattle and sheep) are given in Table 6.1.

McFarland (1971) has shown that the deprivation of food and water results in the build-up of feeding and drinking motivation. Operant investigation of the effects of food restriction on both sows (Lawrence, Appleby & MacLeod, 1988) and broiler breeders (Savory *et al.*, 1993), suggest that their commercial feeding ration (in both cases less than 50% of *ad libitum* intake) result in sustained feeding motivation. Lawrence, Terlouw and Kyriazakis (1993) suggest that this increase in feeding motivation potentiates the animal to forage. Lawrence and Terlouw (1993) suggest that, if an animal that is motivated to forage cannot express this motivation in a species appropriate manner, for example, a hungry pig not being able to root at a suitable substrate such as straw, then it expresses those few

Table 6.1. *Major behavioural and physical restrictions placed on intensively farmed animals*

Species	Class	Restriction
Pigs (*Sus scrofa*)	Fattening	Social distance
		Space
	Breeding	Behavioural contingency
		Nutritional
		Social distance
		Space
Hens (*Gallus gallus domesticus*)	Laying	Behavioural contingency
		Social distance
		Space
	Broilers	Behavioural contingency
		Social distance
		Space
		Water
Cattle (*Bos primigenius taurus*)	Milking	Behavioural contingency
		Maternal care
		Space
	Beef	Behavioural contingency
		Maternal care
	Calves	Maternal care
		Nutritional
Sheep (*Ovis aries*)		Behavioural contingency
		Social distance
		Space

Class in the above Table refers to either the animal's function in a food production system or to its age class. Restriction of social distance means that animals are unable to maintain their preferred distance from conspecifics, this often means they are closer together than they would choose to be as a result of small space allowances. Restriction of behavioural contingency means that when an animal emits an appetitive behaviour (e.g. rooting by a pig) it is not rewarded by the appropriate goal (e.g. a rooting pig does not receive food). Nutritional restriction either means that animals are fed a diet that does not contain all the necessary nutrients for good health or that the amount of food fed is restricted such that the animal is constantly hungry. The restriction of water means that animals are not permitted *ad libitum* access to water. A restriction of maternal care in Table 6.1 means that animals are weaned at a very young age in comparison with natural weaning.
(From Fraser & Broom, 1990.)

components of the behavioural sequence that can be expressed within their environment. Eventually, these behavioural components (e.g. biting, chewing and licking) become 'channelled' into simpler and less complex forms that are thought to give rise to stereotypic behaviour such as bar-biting.

The deprivation of food and water can also have physical effects. For example, food-deprived pigs show a greater instance of gastric lesions because of this stress (Lawrence et al., 1993). It is not just food, water, a mate, opportunity to forage and exercise of which intensively farmed animals are deprived. Other common areas of deprivation include: the opportunity to perform natural pre-partum behaviour, the opportunity to express natural parental behaviour; the opportunity to explore; the opportunity to express control over the environment, and the non-contingency between appetitive behaviour and its end-point (behavioural contingency) (see Table 6.1). Behavioural contingency refers to the situation when an animal emits an appetitive behaviour (goal-seeking behaviour) such as foraging and this behaviour leads to, and is reinforced by, consummatory behaviour (goal-consuming behaviour) such as eating food located by foraging behaviour.

The wild ancestors of domestic farm animals, upon coming across an environment with a high level of physical and behavioural restriction, would probably start exploring for a better environment (Wood-Gush, Stolba & Miller, 1983). The domestic farm animal does not have this option; clearly this may lead to a welfare problem as often indicated by the performance of stereotyped behaviour. An animal's performance of stereotyped behaviours is widely regarded as an indicator of a sub-optimal or recently sub-optimal environment, and is considered by many to indicate poor welfare (e.g. Hughes & Duncan, 1988; Wemelsfelder, 1990; Mason, 1991).

The evidence for 'behavioural needs'

The concept of 'behavioural needs' suggests that the non-performance of certain appetitive behavioural patterns may reduce an animal's welfare. In favour of this concept, Hughes and Duncan (1988) argue that there are cases in which the performance of behaviour has motivationally significant consequences not necessarily related to the animal's functional requirements. They cite cases where (a) animals choose to make an operant response (performing a behaviour in order to receive a reward) to receive when the same food is freely available (contrafreeloading; see Osborne, 1977); (b) animals emit species-typical behavioural responses while performing a learned arbitrary operant response ('misbehaviour'; see Breland & Breland, 1961; Timberlake, 1984); (c) satiated carnivores kill prey (animals perform unnecessary sequences of appetitive behaviour; see Polsky, 1975); (d) exploratory behaviour with no visible end-point (i.e. information gathering; see Wood-Gush and Vestergaard, 1989); and (e) pigeons will peck at a light before food is delivered without the delivery of food being contingent (dependent) on this response (autoshaping; see Bolles, 1979).

The 'behavioural needs' argument has its opponents, Dawkins (1983, 1990) believes that the study of motivation is central to understanding animal welfare and has criticized the 'behavioural needs' idea for not having a sound motivational basis (Dawkins, 1983). Baxter (1983) has argued that, if the end-point of a behaviour can be realized by human modification of the environment, there will be no 'behavioural need' (i.e. it is the end-point of the behaviour that is important to the animal and not the performance of the behaviour). Baxter further advocates environmental design as an animal welfare *panacea*. However, this hypothesis is untestable since only the individual animal can determine when the end-point of a behaviour is adequately provided for. Given that lines of identically selected livestock show considerable individual differences in their preferences for a dust bath (e.g. Petherick, Waddington & Duncan, 1990), this theory, even if correct, cannot be practically used to forward animal welfare (Young, 1993).

The robustness of this evidence supporting the idea of 'behavioural needs', has largely been unchallenged. First, many early contrafreeloading studies reported animals expressing a preference for response-contingent reinforcers (i.e. those reinforcers or rewards obtained by the animal making a response, for example, a rat pressing a lever) over response-independent reinforcers (i.e. freely available reinforcers, for example, food in a bowl; see Jensen, 1963; Singh, 1970; Duncan & Hughes, 1972; Tarte, 1981). These studies are said to provide evidence that the performance of behaviour is rewarding to animals. Osborne (1977) has shown that most cases of this phenomenon can be explained by current learning theory. For example, in many studies of contrafreeloading, animals were only trained to eat food from the operant food source (i.e. unequal training; see Osborne, 1977), and this occurred even when neophobic subjects such as rats were used (Mitchell, Scott & Williams, 1973; Mitchell, Williams & Sutter, 1974). Thus, most studies reporting that animals showed a preference for contra-freeloading until 1977 could be the result of artefacts from the experimental design/paradigm. Studies after this time take into account Osborne's criticisms, often used subjects whose nutritional requirements were poorly understood. For example, Inglis and Ferguson (1986) reported that starlings (*Sturnus vulgaris*) that were not severely hungry preferred to search for food rather than eating the same food freely available from a bowl. However, the type of food that starlings were offered was mealworms (*Tenebrio molitor*) and it is unlikely that meal-worms fully met the starlings' nutritional requirements.

Jensen, Kyriazakis and Lawrence (1993) have shown that domestic pigs fed diets containing insufficient amounts of crude protein will direct forag-ing behaviour towards available substrates such as straw. It is thought that this foraging response reflects pigs searching for a food source that would contain higher levels of protein. Thus, the contrafreeloading of Inglis's and Ferguson's (1986) starlings may have been due to the starlings search for nutrients missing from their diet, rather than due to the reinforcing proper-ties of performing appetitive foraging behaviour. Young (1993) conducted a series of contrafreeloading experiments with domestic pigs, using a nutritionally complete diet and found that pigs would only contrafreeload for 5% of their diet. This low level of contrafreeloading could be explained by pigs periodically sampling the response-contingent food (see Dow & Lea, 1987) to ensure that there was no nutritional benefit of switching to it. Thus, the work of Young (1993) suggests that domestic pigs do not exhibit contrafreeloading to receive the reinforcing effects of performing the behaviour.

Early reports of the phenomenon of 'misbehaviour' were derived from operant training programmes in which animals were taught to perform arbitrary responses to receive food reinforcement. For example, Breland and Breland (1961), for advertising purposes, trained a pig to carry a wooden dime and drop it into a large 'piggy bank' to receive food. After some days of performing the task as trained, the pig rather than carrying the dime would drop it and root it along the ground, delaying the delivery of food reinforcement. Breland and Breland (1961) tried to correct this direction of species-typical behaviour to the dime by increasing the pig's hunger level. However, this only made the pig root the dime even more. Rather than evidence of a 'behavioural need' to perform, foraging behav-iour, or that the performance of foraging behaviour is reinforcing, this evidence merely demonstrates that a hungry pig is motivated to express species-typical foraging behaviour (see Young, 1993). It also demonstrates that it may be difficult for animals to maintain the performance of an arbitrary operant response for reinforcement. This suggestion is supported by Young, MacLeod and Lawrence (1994*b*) who showed that pigs prefer-red to use an operant *manipulandum* (i.e. device to emit responses on to obtain a reward, for example, a lever for a rat to press in order to obtain food) to obtain food that accommodated species-typical foraging behav-iour over an operant *manipulandum* of arbitrary design.

Kruuk and Turner (1967) have observed a lioness (*Panthera leo*) kill up to four Thomson's gazelle (*Gazella thomsoni*) in one morning; however, they suggest that this is not 'surplus' or functionless killing, since lions will stay with their kill for days before consuming it. Leopards (*Panthera pardus*) are known to store food up trees for later consumption and may therefore kill opportunistically even when apparently satiated (not hungry; Kruuk & Turner, 1967). For example, hyena (*Crocuta crocuta*) and leopards engage in surplus killing on very dark nights when prey are unable to flee (Kruuk, 1972). Well-fed domestic cats (*Felis catus*) often catch prey (see Sainsbury, Bennett & Kirkwood, 1995) the purpose of this behaviour may be to practice hunting skills. The suggestion that cats are highly motivated to hunt is made by a number of authors (e.g. Eaton, 1972; Leyhausen, 1979); however, it has never to my knowledge been substantiated by experimental investigation. In fact, cats are one of the few species reported not to contrafreeload (Koffer & Coulson, 1971). Thus I would suggest that the killing of prey by satiated carnivores is not strong evidence that animals have a need to perform unnecessary sequences of behaviour.

It is difficult to prove that animals unable to perform all behavioural components of an appetitive behavioural sequence necessarily suffer from such a limited expression of a behaviour pattern. For example, although it has been demonstrated that domestic hens preferred to build a new nest even when the one they built the previous day is still available (Hughes, Duncan & Brown, 1989); this is not necessarily evidence that hens find nest building reinforcing. One alternative explanation is that, when hens have been disturbed by humans (a potential predator?), they build new nests (i.e. this is a functional response to disturbance by a potential predator).

Strongly related to unnecessary sequences of appetitive behaviour is exploratory behaviour. Berlyne (1960) has said that extrinsic exploration (the search for conventional reinforcers, such as food) is synonymous with appetitive behaviour. It is often impossible to determine the goal of appetitive behaviour until the consummatory phase begins. It has been suggested that animals may acquire information useful for one motivational system by latent learning whilst performing the appetitive behaviour of another (Wood-Gush & Vestergaard, 1989). Intrinsic exploration is directed at stimuli, which may have no biological significance and has the goal of gathering information, possibly for future use (Berlyne, 1960). Thus, sequences of apparently functionless appetitive behaviour may have a function in information gathering.

Rats (*Rattus norvegicus*), for example, in a Y-maze always enter the arm leading to a novel area rather than alternating arms (Barnett, 1963), implying that exploring novel areas is rewarding. Data suggest that it is novelty rather than complexity which augments exploration (Schneider & Gross, 1965). However, exploratory behaviour may be reinforced by sensory stimulation. Barnes and Baron (1961) demonstrated that rats are more motivated to look at complex shapes than simple ones. Additionally, it has been reported that rats cease to contrafreeload (i.e. stop choosing to eat food that requires them to make a response to obtain it rather than eating identical freely available food) when the stimulus change associated with the delivery of operant food is eliminated (e.g. sound or sight of food delivery; see Osborne & Shelby, 1975). This observation suggests that stimulus change acts as a form of secondary reinforcement, and combined with primary reinforcement from the food causes the animal to contrafreeload (Osborne, 1977). Young (1993) reported that an olfactory and taste stimulus change increased contrafreeloading in domestic pigs.

Autoshaping (the performance of non-response contingent appetitive behaviour; that is, operant behaviour not dependent upon receiving a reward for it to be maintained); again there is no direct evidence that the performance of autoshaping is either a 'behavioural need' or reinforcing in itself. It is mistaken to think that the performance of any behavioural pattern has inherent and perceptibly reinforcing properties. Autoshaping is likely to be a type of hard-wired response, such as a reflex (McFarland, 1987). While such a response is beneficial for the subject, it does not have to be reinforcing.

Although there are strong criticisms of the evidence used to support the idea of 'behavioural needs' (that animals have a need to express particular appetitive behaviour in order for their welfare not to be compromised) presented in this chapter, these criticisms do not invalidate the concept. The concept of 'behavioural needs' is not without value to the study of animal welfare since it gives rise to predictions that can be tested. The results of testing any concept can be used to modify, refine or totally refute the concept.

Types of behavioural needs

Certain behavioural patterns, those which restore homeostatic balance, may appear to be a behavioural need, either because the animal is in physiological deficit, or because the performance of the behaviour is important (self-reinforcing) to the animal. For example, an apparent 'behavioural need' for expressing foraging behaviour may result from purely physiological requirements for food, or from a behavioural need to forage. It is important to be able to distinguish which mechanism is responsible and what the consequences are for the animal's welfare.

The application of certain experimental paradigms may elucidate the cause of the 'apparent behavioural need' to forage. For example, foraging animals need food to satisfy their hunger and information about the availability of food (see Krebs & McCleery, 1984). It has been demonstrated that animals are sensitive to variance in food supply (which may be equated to information) and are able to make adaptive foraging decisions based on this information (e.g. risk-sensitivity; see Real & Caraco, 1986). It has been shown that great tits (*Parus major*) perform sampling behaviour to obtain information that allows them to maximize food intake by moving to food patches in the order of highest food density (Smith & Sweatman, 1974). Inglis & Ferguson (1986) demonstrated that 'non-hungry' starlings prefer to search for food rather than eat the same food freely available. In many of the contrafreeloading studies reviewed by Osborne (1977), free food is consumed at the beginning of the experiment and operant food later.

These contrafreeloading studies suggest that, once a certain level of nutrient intake has been achieved, food acquisition may shift from being dependent on physiological need to being more governed by requirements for information (Woodworth, 1958; Inglis, 1983). Thus, foraging behaviour may be motivated by two mechanisms, one to restore homeostatic balance (a physiological need; Woodworth, 1958; Inglis & Ferguson, 1986) and the other to gain information about the environment (an information need; Woodworth, 1958; Inglis & Ferguson, 1986).

It may be possible to use a risk-sensitive foraging paradigm to determine which 'need' an animal may be showing. Animals in high energy deficit (i.e. hungry) given the choice of two food patches, one with high variance in reward rate (variable site) and the other with no variance (constant site), usually choose the variable site (risk prone behaviour, i.e. the animal risks obtaining no food by gambling on receiving a lot of food) because by doing so they may obtain enough food to survive, whereas the constant site cannot provide enough food for survival. Conversely, animals with little energy deficit may prefer either the constant site (risk adverse behaviour, i.e. the animal is choosing to receive a constant amount of food) or neither site (risk indifference behaviour, i.e. the animal expresses no preference for constant or variable sources of food). It could be argued that animals showing risk prone behaviours are demonstrating foraging behaviour motivated by a physiological need (see Woodworth, 1958; Inglis, 1983; Inglis & Ferguson, 1986). With these methods, it may be possible to differentiate between homeostatically motivated foraging behaviour (physiological need) and foraging behaviour that may be self-reinforcing (information need).

Behavioural need or behavioural void

The issue of whether an animal needs to express certain behavioural patterns to maintain a good state of welfare is difficult to answer (see Mason & Mendl, 1993). The consequences of many behavioural patterns directly affect an animal's welfare, such as drinking, eating, nest-building, grooming, etc. However, it is more difficult to answer whether the animal's welfare is reduced by not being able to perform the behavioural pattern itself (see section 'The evidence for "behavioural needs"').

It may be that what is important is the performance of a behaviour to fill the behavioural void left by, for example, not performing foraging behaviour. Studies of wild pigs show that they may spend up to 75% of their waking time foraging (Stolba & Wood-Gush, 1989). Hughes and Duncan (1988) suggest that, while wild animals carefully budget their time to survive, the intensively farmed animal's problem is to fill the time available from a limited number of behaviours open to it. Broom (1987) has noted that intensively farmed animals have no need to find food, water or shelter as it is provided by their human caretakers. Wood-Gush and Vestergaard (1989) have suggested that, given the opportunity, once homeostatic balance is attained, neophilic species (species that are attracted to new or novel things) use their time to perform inquisitive exploration and neophobic species (species that avoid new or novel things) show appetitive behaviour of the least satisfied motivation.

Many scientists rightly emphasize the importance of providing animals with the opportunity to perform behavioural patterns with inelastic demand functions (i.e. behaviours which animals regard as necessities) because animals are highly motivated to express such behaviours (e.g. feeding; Dawkins, 1983, 1990; Matthews & Ladewig, 1994). However, Hughes and Duncan (1988) suggest that behavioural patterns with the ability to expand and fill a behavioural void created by barren environments, with little behavioural opportunities, are also important for animal welfare. Behaviours that can expand to fill a behavioural void will, by their nature, have an elastic demand function. Thus, while scientists may place a great emphasis on allowing animals to express behavioural patterns with inelastic demand functions (e.g. feeding; Dawkins, 1990; Matthews & Ladewig, 1994) to ensure good animal welfare, certain behavioural patterns with elastic demand functions may also prove to be important for animal welfare (Hughes & Duncan, 1988). A candidate behavioural pattern to fill a behavioural void is intrinsic exploration (i.e. information gathering; Berlyne, 1960; Wood-Gush & Vestergaard, 1989), since it does not have an immediate biological significance (Berlyne, 1960), but can none the less usefully fill time.

The 'behavioural need' argument is further undermined by the flexibility of behaviour demonstrated by many species. Examples of flexible behaviour patterns include positive social interactions of group housed solitary species (e.g. orang utans, *Pongo pygmaeus*; Poole, 1987); modification of feeding and foraging behaviour (e.g. during the war in Vietnam tigers, *Panthera tigris*, scavenged the human bodies rather than hunting; Jackson, 1992), and modification of territorial displays (e.g. fan-tailed warblers, *Cisticola juncidis*, emit their territorial song from artificial structures, such as telegraph poles, rather than flying around their territory; MacGregor *et al.*, 1990). Thus, many behavioural patterns are adaptive and flexible.

If specific behavioural needs do exist, then they may occur in conjunction with behavioural patterns that are more fixed in nature, such as fixed action patterns, operant or innate behaviours. Such behavioural patterns are often 'hard-wired' which are inherently limited in their flexibility. A good example is the egg retrieval response of greylag geese (*Anser anser*). If the egg of a greylag goose rolls out from beneath a brooding bird, then the bird reaches out with its head and retrieves the egg by rolling it back into the nest. If an egg that has rolled out is removed by an experimenter, then the goose continues to perform the retrieval response over and over again, despite the fact that there is no egg to retrieve. Thus the bird is stuck within a pre-programmed behavioural loop.

The implications of fixed actions patterns or limited flexibility of behaviour for animal welfare is that an organism whose behaviour patterns are largely fixed or inflexible in nature will require an environment that allows it to perform more of its wild-type behaviour than a more flexible organism (Skelton, 1994). In practical terms this means that the 'behavioural needs' of simpler organisms such as birds are less likely to be fulfilled by substitute or alternative behaviours. Thus the 'need' of a domestic hen cannot be satisfied by performing an alternative behaviour, whereas it may be possible to fulfil the 'need' of a pig by allowing it to perform an alternative behaviour. In terms of environmental enrichment, it means that the environments of simpler animals are best enriched by simulating the wild environment, whereas while this approach would work with more complex animals, it is not necessary. For example, captive chimpanzees (*Pan troglo-*

dytes) can derive sensory and mental stimulation from working at computer games and challenges (Matsuzawa, 1989), as well as from living in a complex 'wild' environment.

Methods of assessing psychological well-being

Behavioural economics

Optimality theory is widely applied to many aspects of animal behaviour (Krebs & Davies, 1993). Fundamental to the application of optimality theory is a cost–benefit analysis of the animal's behaviour. However, a cost–benefit analysis is only useful if an animal has a limited 'budget' to 'spend from'. The currency most widely used by behavioural ecologists is energy (e.g. Stephens, 1981). However, Dawkins (1990) believes that the budgeting of time may be more important to the animal. The behavioural decisions or choices made by animals are therefore, by analogy, thought to be similar to the economic choices made by a human consumer with a limited financial budget. Thus, products bought under varying degrees of financial pressure are expected to reflect the relative importance of the product to the consumer.

Dawkins (1983) showed that consumer demand theory could be applied to domestic hens, by making them pay an energy cost to get access to more space or conspecifics. Currently, some animal welfare scientists believe that the application of consumer demand theory coupled with the assessment of motivation by operant conditioning techniques is the best way of assessing animal welfare (Hughes & Duncan, 1988; Dawkins, 1990). However, consumer demand theory and operant conditioning techniques are not a *panacea* for assessing behavioural needs.

Consumer demand theory fails to address that animals have both a limited energy and time budget. Most consumers 'buy' many products in a relatively short time scale. The number of products and the quantity of each product bought is dependent upon the size of the budget. Houston and McFarland (1981) have used the idea of 'behavioural resilience', whereby the consumption of any one product is compared to all other products bought. In behavioural terms, the ease with which a behaviour is compressed in time and its proportional change in the time budget is referred to as its behavioural resilience. In reality, all experiments measuring the demand function of reinforcers (i.e. applying consumer demand theory) are a sub-component of the process of measuring behaviour resilience.

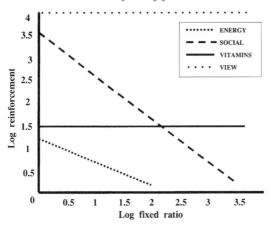

Fig. 6.1 Shapes of demand curves. The *y*-axis represents how much of a particular reinforcer or product an animal has 'worked' (responded) to obtain. The *x*-axis represents the fixed ratio used. The products measured are energy, vitamins, view (being able to look into an empty pen), and social (being able to enter a room with a conspecific inside it).

Demand functions

If the price of a product goes up and the consumer continues to buy the same amount of the product, then this product is said to show an 'inelastic' demand function. Conversely, if when the price of a product goes up the consumer buys less of the product, then this product is said to show an 'elastic' demand function. Products showing inelastic demand functions are usually necessities such as bread, whereas products showing elastic demand functions are often luxuries such as ice-cream (Lea, 1978).

Demand elasticity can be quantified by measuring the slope of the function relating quantity bought (Y) to price (on log–log coordinates; Lea, 1978; see Fig. 6.1). The function is usually of the form:

$$Y = aX + b \tag{1}$$

The absolute value of the normally negative slope (a) is the elasticity coefficient and (b) is a constant. Thus, curves with an elasticity value of less than 1.0 are described as elastic. The elasticity of demand is influenced by income, availability of substitutable products, and the nature of the commodity (Lea, 1978).

It has been suggested that the amount of behavioural effort required to obtain a resource may be equivalent to price (Lea, 1978). The level of effort required may be increased systematically by conditioning an animal to press a key to obtain the desired resource and then increasing the number of key presses for reinforcement. Operant conditioning techniques as originally devised by Skinner (1932) for studying learning can be used for this purpose. The simplest experimental protocol is to increase the behavioural effort required by using increasing fixed ratio schedules of reinforcement (this is when an animal receives a food reward for each sequence of five key presses, for example).

Fig. 6.1 shows curves (slopes) typical for this type of experimental paradigm, that is commodities such as food (energy and vitamins) showing relatively inelastic demand (i.e. little or no slope), social contact and the opportunity to view an empty pen, elastic demand (i.e. quite steeply sloped; Dawkins, 1983, 1990; Matthews & Ladewig, 1994). Most scientists believe that it is the slope of the demand curve that highlights what is most important to animals, since this reflects the elasticity of demand, for example, necessities have an inelastic demand (e.g. bread) and luxuries have an elastic demand (e.g. ice-cream; Dawkins, 1990; Matthews & Ladewig, 1994). However, Ng (1990) has argued that it is the area bounded by the curve (the 'demand area') that is important for assessing animal welfare. The area bounded by the curve highlights how much of a commodity an animal will 'buy' (i.e. consume) but suggests nothing about how important the product is to the animal.

Clearly, the amount of a product bought is important to an animal, for example, animals have requirements to consume certain levels of nutrient per day in order to survive. An adult pig, for example, will require several hundred units of energy per day, but only a few grams of vitamins. Thus, the amounts of different products 'bought' by an animal will reflect its physiological requirement. If we consider energy and vitamins in Fig. 6.1, taking the area bounded by the curve as our measure of what is important to animals, we have to conclude that energy is more important than vitamins, since energy has a larger 'demand area' (area bounded by the curve). Furthermore, if we compare 'vitamins' and 'social' (social contact) in Fig. 6.1 using the 'demand area', we conclude that 'social' is more important to the animal than 'vitamins'. While social contact for a social species is obviously important, it is probably not as important as 'vitamins', since appropriate level of vitamin consumption affect an animal's immediate chances of survival. If, instead of using 'demand area' as our measure of what is important to animals, we use slope of the demand curve, with slopes closest to 1.0 (i.e. inelastic demand curves) being indicating what is most important to animals. Then the animal's priority order becomes: first, 'energy' and 'vitamins' and, second 'social' and 'view'.

Thus, the analysis of demand curves (slopes) produces a more logical ordering of priorities for animals than the use of demand areas.

Dawkins (1983, 1990) has suggested that the demand curve produced by food for an animal is a useful baseline for comparing with other products (reinforcers), since it is essential for survival and is therefore likely to show an inelastic demand function (see Matthews & Ladewig, 1994). Thus, a hierarchy of behavioural needs can be established by using food as a baseline for comparison.

The wide acceptance of these techniques by animal welfare scientists (e.g. Dawkins, 1990) because it produces easily quantifiable results (i.e. slope values) has resulted in a 'shopping list' approach to 'behavioural needs'; that is, rather than considering the overall welfare of an animal, its welfare is now being divided up into lists of behavioural patterns that 'need' or do not 'need' to be expressed (see Jensen & Toates, 1993). There are many problems involved with the application of Consumer Demand Theory; however, due to space limitations, only a few are listed here. These problems are first, that the experiments are time consuming and must be repeated for every behavioural pattern and resource, and for every species. Also, it is difficult to make comparisons between an animal's demand for different recourses. Although some scientists use the slope of the demand function to measure elasticity (the sensitivity of a commodity to variations in price), experimental differences make direct comparisons difficult. For example, few studies report the effort required to make the operant response, and yet Armus (1986) has shown that, the greater the response effort required, the fewer operant responses are made. The variance introduced to the results by using different levels of effort to make an operant response is likely to be compounded by the open economy experimental designs used in such studies (where an animal's food intake, or other reinforcers, is partially determined by the experimenter; that is, animals are only briefly in the experimental set-up and receive food outside of the experimental set-up; see Houston & McNamara, 1989) used in such studies. For example, in Matthew's and Ladewig's (1994) study, pigs were fed in home pens at 0700 and 1600 hours, and then placed in the experimental set-up for 100 minutes of testing. The adoption of closed economy design of experiments, where the animal essentially lives in the experiment and receives all of its food, or other reinforcers, in the experiment (in this case by operant responding) may reduce the variation introduced by inter-experimental differences in consumer demand studies since closed economy design of experiments is more likely to reflect what is important to animals in the long term.

Limitations of operant conditioning techniques

In many studies where consumer demand theory is applied, operant conditioning techniques are used to make the animals pay a cost to express a behaviour or obtain a resource. Typically, animals are trained to emit responses, such as panel pressing by pigs, for food. Animals often take a long time to learn to make such responses (Petherick & Rutter, 1991), making the practical application of this technique limited. The question is whether having a pig press a panel is really imposing any kind of energy cost on the animal, especially given that, in many operant conditioning experiments, touch sensitive panels are used (the effort required to operate such panels is often negligible). Hutson (1991) has suggested that the level of energy expended by sows responding operantly for food is equivalent to the level of energy expended by a sow stereotyping (11 kJ/kg live weight$^{0.75}$ per hour; see Cronin et al., 1986). However, Hutson's suggestion is not based on metabolic evidence and is therefore, at best, speculation. Although it is unlikely that operating a touch-sensitive panel is imposing much of an energy cost on an animal, the time required to complete a response sequence (i.e. say five presses for a food reward) is obviously imposing a time cost. Obviously, the greater the number of responses an animal is required to make in order to receive a reinforcer (e.g. food), the greater the time cost it incurs.

Young, MacLeod and Lawrence (1994b) have demonstrated that the level of operant responding is affected by the design of the *manipulandum* (the mechanical device upon which an animal directs an operant response in order to obtain a reinforcer, for example, the 'lever' a rat presses in a Skinner box in order to obtain food). Specifically, they demonstrated that animals make more operant responses and thus receive a higher level of reinforcement when the design of the *manipulandum* accommodates the appropriate species-typical behaviour for a specific reinforcer (in this case food). The implications of their findings are that the absolute level of an animal's motivation may be under-estimated by *manipulanda* that do not accommodate species-specific behaviour. An explanation for this finding is species-specific constraints on learning (i.e. animals cannot learn all pairings of response and reinforcer with equal ease; see Roper, 1983).

Matthews and Ladewig (1994) used operant conditioning techniques in order to assess the behavioural demand functions of pigs for food, social companionship, and the opportunity to view an empty pen. They found that pigs showed an inelastic demand function for food, an elastic demand function for social companionship, and a more elastic demand function for the opportunity to see an empty pen. Applying consumer demand theory to their results, they derived a priority order of first food, then social companionship, and finally the opportunity to see an empty pen. However, it could be argued that their results merely reflect the degree to which their *manipulandum* accommodated the species-specific behavioural pattern to acquire the respective reinforcer. For example, if the *manipulandum* was of such a design that the pigs could easily direct a foraging type behaviour towards it, then a higher level of responses for this behaviour would be expected when compared with the level of operant responding for access to opportunities to express social behaviour. It should be noted that many operant *manipulanda* were originally developed to measure feeding behaviour (e.g. the Skinner box, Skinner, 1932) and are now used to measure how an animal will work for a variety of reinforcers (e.g. social contact, heat, light and so on; see, for example Matthews & Ladewig, 1994). Bolles (1988) goes so far as to argue that operant *manipulanda* are not the arbitrary design that psychologists claim (see Roper, 1986) but have primarily evolved around the feeding behaviour of the experimenter's subjects. Thus it is not surprising that they accommodate feeding behaviour more than social behaviour.

Although some authors have suggested that the slope of the demand function of food can be used as a baseline for comparison with other reinforcers (Dawkins, 1990; Matthews & Ladewig, 1994), experimental differences often make such comparisons of limited value. For example, Young, MacLeod and Lawrence (1994*b*) demonstrated that the level of operant responding made by an animal was dependent upon the design of the *manipulandum*. Thus, an animal may appear to show no motivation to express a certain behaviour because of the incompatibility between the animal's behaviour and the design of the *manipulandum*. For example, Dawkins and Beardsley (1986) were unable to condition domestic hens to make an operant response (e.g. key pecking or operating a floor treadle with their foot) in order to have access to a dust bath. Beilharz and Zeeb (1981) would interpret the inability of hens to learn to make an operant response to obtain a dust bath as indicating that hens are not highly motivated to perform dust bathing (i.e. hens do not have a behavioural need to dust bath). However, Smith, Appleby and Hughes (1990) found that hens could learn to open one-way doors the wrong way to gain access to a dust bath, and would spend on average 34.4 minutes trying to open the door to the dust bath before succeeding. The results of this study therefore suggest that hens are highly motivated to work for access to a dust bath. Thus, the (physical) form of the operant response that an animal is required to make for a specific reinforcer can affect the level of responding and how important animal welfare scientists' believe that reinforcer to be to the animal.

Behavioural resilience

In reality, the application of consumer demand theory, coupled with the use of operant conditioning procedures, simultaneously varies energy and time costs (since expressing behaviour has an energetic cost and uses up time). One method of avoiding this problem is to measure behavioural resilience. The behavioural resilience of a behaviour pattern is the degree to which a behaviour can resist being compressed in the time budget of an animal that has been forced to spend more of its time budget on a certain behaviour such as foraging. Behavioural resilience can be measured simply by significantly increasing an animal's foraging time by diluting its food with an inert and inedible substrate (e.g. Lemon & Barth, 1992) and measuring changes in the resulting budgeting of time. Since the animal is now spending more time foraging, it has less time to perform other behaviours. Therefore, it can be assumed that those behaviours that have become compressed in time are less important. For example, gelada baboons, *Theropithecus gelada*, with an increased demand for food (e.g. lactating females) spend more time foraging, an equal amount of time socializing, and less time resting than those baboons with a smaller demand for food (Dunbar & Dunbar, 1988). Thus, time-stressed baboons re-organize their time budgets in a hierarchy of importance. Swennen, Leopold and DeBruijn (1989) and Young (1993) have demonstrated similar effects for time stressed oyster catchers, *Haematopus ostralegus*, and pigs. This technique is not only more time-effective than operant conditioning and consumer demand theory techniques, but also produces a hierarchy of behavioural needs (see Young, 1993) in one experiment.

Conclusions

To conclude, it is apparent that more complex animals (e.g. mammals) rarely suffer from the inability to express a particular behavioural pattern. However, the 'behavioural void' created by being in a captive and possibly barren environment with little to occupy an animal's time may be a welfare problem in that the animal must find ways in which to fill this void. Contrastingly, simpler animals (e.g. birds) are more likely to suffer reduced welfare from the non-performance of a particular behaviour. In practice, it may prove impossible to distinguish between welfare problems arising from the inability to express specific natural behavioural patterns and those resulting from a behavioural void.

In animal welfare terms, 'hard-wired' and homeostatically motivated behaviour should be accommodated for in a species-typical manner. Breland and Breland (1961) failed to teach their 'organisms' species-typical behaviours that were linked to the appropriate reinforcer which resulted in 'misbehaviour' by these animals. Lawrence and Terlouw (1993) have argued that, if intensively housed pigs are unable to express their commercially induced hunger in species-typical foraging behaviour, then it may become channelled into simpler forms, which eventually develop into stereotypies. Young, Carruthers and Lawrence (1994a) have tackled this problem for intensively farmed pigs, by producing a foraging device, 'The Edinburgh Foodball', which allows hungry pigs to direct complex and variable species-typical foraging behaviour (e.g. rooting) towards it. The Foodball comprises a spherical casing with a food dispensing hole and an internal food store which is fitted with a meterable food dispensing mechanism. Food is delivered to a pig randomly in time, space and quantity, thereby mimicking the distribution of food under natural conditions, in response to the pig rooting the Foodball. Unfortunately, it appears that enrichment devices such as the Foodball are unlikely to be used in commercial practice because of the additional cost of buying them and the increase in human labour involved in implementing them.

An animal's need to perform non-homeostatically (physiologically) motivated foraging behaviour may be satisfied by occupational activities that share common appetitive sequences with foraging behaviour, intrinsic exploration, for example, Wood-Gush and Vestergaard (1991) found that providing pigs with novel objects that were changed frequently stimulated exploratory behaviour (2.3–92 times longer than when pigs were presented with familiar objects) and play behaviour. The exploration of novel objects may therefore be an alternative behaviour that domestic pigs could be given the opportunity to perform when their foraging behaviour is governed by non-homeostatic needs (e.g. the need for information).

In terms of experiments I would like to suggest three ways of improving the way in which we assess the welfare of animals: (a) the use of 'behavioural resilience' rather than 'consumer demand theory' approach; (b) if operant *manipulanda* are to be used, then they should be designed around species-species behaviour; and (c) animals are much more likely to 'tell' the experimenter what is important to them if they are maintained in a closed economy experimental design.

Finally, to enrich the lives of complex animals such as pigs, it is not necessary to simulate the 'wild' merely to stimulate their senses and their brain. However, for simpler animals such as hens, the only way to stimulate their senses and their brain may be by simulating the 'wild'.

Acknowledgements

This chapter has greatly benefited from my numerous conversations with the members of the G.A.B.S. 'Behaviour team' over the last 5 years. I would especially like to thank Alistair Lawrence, Ilias Kyriazakis, Mike Mendl and Françoise Wemelsfelder. I would also like to thank Francine Dolins and two anonymous referees for their helpful comments. The Royal Zoological Society of Scotland graciously supported my work in producing this chapter. Finally, I would like to thank Beverley Williams for typing the references and for proof-reading.

References

Armus, H.L. (1986). Effect of response effort requirement on frequency of short interresponse times. *Bulletin of the Psychonomic Society*, **24**, 284–5.

Barnes, G. W. & Baron, A. (1961). Stimulus complexity and sensory reinforcement. *Journal of Comparative and Physiological Psychology*, **54**, 466–9.

Barnett, S. A. (1963). *The Rat: A Study in Behavior*. London: Methuen.

Baxter, M. R. (1983). Ethology in environmental design for animal production. *Applied Animal Ethology*, **9**, 207–220.

Beilharz, R. G. & Zeeb, K. (1981). Applied ethology and animal welfare. *Applied Animal Ethology*, **7**, 3–10.

Berlyne, D. E. (1960). *Conflict, Arousal, and Curiosity*. New York: McGraw-Hill.

Bolles, R. C. (1979). *Learning Theory*. 2nd edn, pp. 180–3. New York: Holt, Reinhart and Winston.

Bolles, R. C. (1988). The bathwater and everything. *Behavioral and Brain Sciences*, **11**, 429–93.

Breland, K. & Breland, M. (1961). The misbehaviour of organisms. *American Psychologist*, **16**, 681–4.

Broom, D. M. (1987). Farm animal behaviour. In *The Oxford Companion to Animal Behaviour*, ed. D. McFarland, pp. 176–80. Oxford: Oxford University Press.

Command Paper 2836 (1965). *Report of the Technical Committee to Enquire into the Welfare of Animals Kept under Intensive Livestock Husbandry Systems*. London: HMSO.

Cronin, G. M. , Tartwijk, J. M. F. M. van, Hel, W. van der & Verstegen, M. W. A. (1986). The influence of degree of adaptation to tethering by sows in relation to behaviour and energy metabolism. *Animal Production*, **42**, 257–68.

Dawkins, M. S. (1983). Battery hens name their price: consumer demand theory and the measurement of ethological needs. *Animal Behaviour*, **31**, 1195–205.

Dawkins, M. S. (1990). Form an animal's point of view: motivation, fitness, and animal welfare. *Behavioral and Brain Sciences*, **13**, 1–61.

Dawkins, M. S. & Beardsley, T. (1986). Reinforcing properties of access to litter in hens. *Applied Animal Behaviour Science*, **15**, 351–64.

Dow, S. M. & Lea, S. E. G. (1987). Sampling of schedule parameters by pigeons: Tests of optimizing theory. *Animal Behaviour*, **35**, 102–14.

Dunbar, R. I. M. & Dunbar, P. (1988). Maternal time budgets of gelada baboons. *Animal Behaviour*, **36**, 970–80.

Duncan, I. J. H. & Hughes, B. O. (1972). Free and operant feeding in domestic fowls. *Animal Behaviour*, **20**, 775–7.

Eaton, R. L. (1972). An experimental study of predatory behaviour and feeding behaviour in the cheetah (*Acinonyx jubatus*). *Zeitschrift fur Tierpsychology*, **31**, 270–80.

Fraser, A. F. & Broom, D. M. (1990). *Farm Animal Behaviour and Welfare*. 3rd edn. London: Baillière Tindall.

Glickman, S. E. & Schiff, B. B. (1967). A biological theory of reinforcement. *Psychological Review*, **74**, 81–109.

Herrnstein, R. J. (1977). The evolution of behaviourism. *American Psychologist*, **32**, 593–603.

Houston, A. I. & McFarland, D. J. (1981). Behavioural resilience and its relation to demand functions. In *Limits to Action: The Allocation of Individual Behaviour*. ed. J. E. R. Staddon, pp. 177–203. New York: Academic Press.

Houston, A. I. & McNamara, J. M. (1989). The value of food: effects of open and closed economies. *Animal Behaviour*, **37**, 546–62.

Hughes, B. O. & Duncan, I. J. H. (1988). The notion of ethological 'need', models of motivation and animal welfare. *Animal Behaviour*, **36**, 1696–707.

Hughes, B. O. , Duncan, I. J. H. & Brown, M. F. (1989). The performance of nest-building by domestic hens: is it more important than the construction of a nest. *Animal Behaviour*, **37**, 210–14.

Hutson, G. D. (1991). A note on hunger in the pig: sows on restricted rations will sustain an energy deficit to gain additional food. *Animal Production*, **52**, 233–5.

Inglis, I. R. (1983). Towards a cognitive theory of exploratory behaviour. In *Exploration in Animals and Humans*, ed. J. Archer & L. Birke, pp. 72–116. Wokingham: Van Nostrand Reinhold.

Inglis, I. R. & Ferguson, N. J. K. (1986). Starlings search for food rather than eat freely available, identical food. *Animal Behaviour*, **34**, 614–16.

Jackson, P. (1992). *Endangered Species: Tigers*. London: Chartwell Books.

Jensen, G. D. (1963). Preference for bar pressing over 'freeloading' as a function of number of rewarded presses. *Journal of Experimental Psychology*, **65**, 451–4.

Jensen, M. B. , Kyriazakis, I. & Lawrence, A. B. (1993). The activity and straw directed behaviour of pigs offered foods of different protein content. *Applied Animal Behaviour Science*, **37**, 211–21.

Jensen, P. & Toates, F. (1993). Who needs 'behavioural needs'? Motivational aspects of the needs of animals. *Applied Animal Behaviour Science*, **37**, 161–81.

Koffer, K. & Coulson, G. (1971). Feline indolence: cats prefer free to response-produced food. *Psychonomic Science*, **24**, 41–2.

Krebs, J. R. & Davies, N. B. (1993). *An Introduction to Behavioural Ecology*, 3rd edn. Oxford: Blackwell Scientific Publications.

Krebs, J. R. & McCleery, R. H. (1984). Optimization in behavioural ecology. In *Behavioural Ecology: An Ecological Approach*, ed. J. R. Krebs & N. B. Davies, Chapter 4, pp. 91–121. Oxford: Blackwell Scientific Publications.

Kruuk, H. (1972). Surplus killing by carnivores. *Journal of Zoology*, **166**, 233–44.

Kruuk, H. & Turner, M. (1967). Comparative notes on predation by lion, leopard, cheetah and wild dog in the Serengeti area, East Africa. *Mammalia*, **31**, 1–27.

Lawrence, A. B. & Terlouw, E. M. C. (1993). Behavioural factors involved in the development and continued performance of stereotypic behaviours. *Journal of Animal Science*, **71**, 2815–25.

Lawrence, A. B. , Appleby, M. C. & MacLeod, H. A. (1988). Measuring hunger in pigs using operant conditioning: the effect of food restriction. *Animal Production*, **47**, 213–20.

Lawrence, A. B. , Terlouw, E. M. C. & Kyriazakis, I. (1993). The behavioural effects of undernutrition in confined farm animals. *Proceedings of the Nutrition Society*, **52**, 219–29.

Lea, S. E. G. (1978). The psychology of economics and demand. *Psychology Bulletin*, **85**, 441–66.

Lemon, W. C. & Barth, R. H. (1992). The effects of feeding rate on reproductive success in the zebra finch, *Taeniopygia guttata*. *Animal Behaviour*, **44**, 851–7.

Leyhausen. P. (1979). *Cat Behaviour*. New York: Garland Inc.

MacGregor, P. K. , Clayton, H. S. , Kolb, U. , Stockley, P. & Young, R. J. (1990). Individual differences in the displays of Fan-tailed Warblers *Cisticola juncidis*: associations with territory and male quality. *IBIS*, **132**, 111–32.

MAFF (1983). *Codes of Recommendations for the Welfare of Livestock: Cattle*. London: Ministry of Agriculture, Fisheries, and Food.

Mason, G. (1991). Stereotypies: a critical review. *Animal Behaviour*, **41**, 1015–37.

Mason, G. & Mendl, M. T. (1993). Why is there no simple way of measuring animal welfare? *Animal Welfare*, **2**, 301–19.

Matsuzawa, T. (1989). Spontaneous pattern construction in a chimpanzee. In *Understanding Chimpanzees*, ed. P. Heltne & L. Marquardt. Cambridge, Massachusetts: Harvard University Press.

Matthews, L. R. & Ladewig, J. (1994). Environmental requirements of pigs measured by behavioural demand functions. *Animal Behaviour*, **47**, 713–19.

McFarland, D. J. (1971). *Feedback Mechanisms in Animal Behaviour*. London: Academic Press.

McFarland, D. J. (1985). *Animal Behaviour: Psychobiology, Ethology and Evolution*. Bath: Longman, Scientific and Technical.

McFarland, D. J. (1987). Reflex. In *The Oxford Companion to Animal Behaviour*, ed. D. McFarland, pp. 470–1. Oxford: Oxford University Press.

Mitchell, D. , Scott, D. W. & Williams, K. D. (1973). Container neophobia and the rats preference for earned food. *Behavioural Biology*, **9**, 613–24.

Mitchell, D. , Williams, K. D. & Sutter, J. (1974). Container neophobia as a predictor of preference for earned food by rats. *Bulletin of the Psychonomic Society*, **4**, 182–4.

Ng, Y-K. (1990). The case for and using 'demand areas' to measure changes in well-being. *Behavioral and Brain Sciences*, **13**, 30–1.

Osborne, S. R. (1977). The free food (contrafreeloading) phenomenon: a review and analysis. *Animal Learning and Behaviour*, **5**, 221–35.

Osborne, S. R. & Shelby, M. (1975). Stimulus change as a factor in response maintenance with free food available. *Journal of the Experimental Analysis of Behaviour*, **24**, 17–21.

Petherick, J. C. & Rutter, S. M. (1991). Quantifying motivation using a computer-controlled push door. *Applied Animal Behaviour Science*, **27**, 159–67.

Petherick, J. C. , Waddington, D. & Duncan, I. J. H. (1990). Learning to gain access to a foraging and dustbathing substrate by domestic fowl: 'is out of sight out of mind?' *Behavioural Processes*, **22**, 213–26.

Polsky, R. H. (1975). Hunger, prey feeding and predatory aggression. *Behavioural Biology*, **13**, 81–93.

Poole, T. B. (1987). Social behaviour of a group of orang utans (*Pongo pygmaeus*) on an artificial island in Singapore Zoological Gardens. *Zoo Biology*, **6**, 315–30.

Poole, T. B. (1992). The nature and evolution of behavioural needs in mammals. *Animal Welfare*, **1**, 203–20.

Real, L. & Caraco, T. (1986). Risk and foraging in stochastic environments. *Annual Review of Ecology and Systematics*, **17**, 371–90.

Roper, T. J. (1983). Learning as a biological phenomenon. In *Animal Behaviour*, vol. 3, *Genes, Development and Learning*, ed. T. R. Halliday & P. J. B. Slater, Chapter 6 pp. 178–212. Oxford: Blackwell Scientific Publications.

Sainsbury, A. W. , Bennett, P. M. & Kirkwood, J. K. (1995). The welfare of free-living wild animals in Europe: harm caused by human activities. *Animal Welfare*, **4**, 183–206.

Savory, C. J. , Maros, K. & Rutter, S. M. (1993). Assessment of hunger in growing broiler breeders in relation to a commercial restricted feeding programme. *Animal Welfare*, **2**, 131–52.

Schneider, G. E. & Gross, C. G. (1965). Curiosity in the hamster. *Journal of Comparative and Physiological Psychology*, **59**, 150–2.

Singh, D. (1970). Preference for bar pressing to obtain reward over freeloading in rats and children. *Journal of Comparative and Physiological Psychology*, **73**, 320–7.

Skelton, T. (1994). Lizard Welfare. Unpublished MSc thesis, University of Edinburgh, UK.

Skinner, B. F. (1932). On the rate of formation of a conditioned reflex. *Journal of General Psychology*, **7**, 274–86.

Smith, J. N. M. & Sweatman, H. P. A. (1974). Food searching behaviour of titmice in patchy environments. *Ecology*, **55**, 1216–32.

Smith, S. F. , Appleby, M. C. & Hughes, B. O. (1990). Problem-solving by hens: opening doors to reach nest sites. *Applied Animal Behaviour Science*, **28**, 287–92.

Stephens, D. W. (1981). The logic of risk-sensitive foraging preferences. *Animal Behaviour*, **29**, 628–9.

Stolba, A. & Wood-Gush, D. G. M. (1989). The behaviour of pigs in a semi-natural environment. *Animal Production*, **48**, 419–25.

Swennen, C. , Leopold, M. F. & DeBruijn, L. L. M. (1989). Time stressed oystercatchers, *Haematopus ostralegus*, can increase their intake rate. *Animal Behaviour*, **38**, 8–22.

Tarte, R. D. (1981). Contrafreeloading in humans. *Psychological Reports*, **49**, 859–66.

Timberlake, W. (1984). A temporal limit on the effect of future food on current performance in an analogue of foraging and welfare. *Journal of the*

Experimental Analysis of Behaviour, **41**, 117–24.

Wemelsfelder, F. (1990). Boredom and laboratory animal welfare. In *The Experimental Animal in Biomedical Research*, ed. B. E. Rollin, Boca Raton: CRC Press.

Whittemore, C. T. (1987). *Elements of Pig Science*. Hong Kong: Longman, Scientific and Technical.

Wood-Gush, D. G. M. & Vestergaard, K. (1989). Exploratory behaviour and the welfare of intensively kept animals. *Journal of Agricultural Ethics*, **2**, 161–9.

Wood-Gush, D. G. M. & Vestergaard, K. (1991). The seeking of novelty and its relation to play. *Animal Behaviour*, **42**, 599–606.

Wood-Gush, D. G. M. , Stolba, A. & Miller, C. (1983). Exploration in farm animals and animal husbandry. In *Exploration in Animals and Humans*, ed. J. Archer & L. Birke, Chapter 8, pp. 198–209. Wokingham: Van Nostrand Reinhold.

Woodworth, R. S. (1958). *Dynamics of Behaviour*. New York: Holt, Reinhart and Winston.

Young, R. J. (1993). Factors affecting foraging motivation in the domestic pig. Unpublished PhD Thesis, University of Edinburgh, UK.

Young, R. J. , Carruthers, J. & Lawrence, A. B. (1994*a*). The effect of foraging device (The 'Edinburgh Foodball') on the behaviour of pigs. *Applied Animal Behaviour Science*, **39**, 237–47.

Young, R. J. , MacLeod, H. A. & Lawrence, A. B. (1994*b*). Effect of manipulandum design on operant responding in pigs. *Animal Behaviour*, **47**, 1488–90.

7

Personality and the happiness of the chimpanzee

JAMES E. KING

Two hunters, Ntino and Iko were out strolling one day through the forest. They came across some chimpanzees who were playing in the branches of a mulemba tree.
'Look at the chimpanzees,' Ntino said. 'Look how they swing so easily through the branches. This is the happiness of the chimpanzee.'
'How can you know?' Iko said. 'You are not a chimpanzee. How can you know if it is happy or not?'
'You are not me,' Ntino said. 'How do you know that I do not know the happiness of the chimpanzee?'
William Boyd
Brazzaville Beach

This dialogue between Ntino and Iko about the happiness of chimpanzees addresses two approaches to a problem that the philosophically inclined have pondered for centuries. The solipsist Iko has no patience with frivolous speculation about chimpanzee happiness. However, the clever Ntino redirects Iko's solipsist argument against Iko himself. Ntino also reveals himself to be an adherent of the increasingly popular view that animals are most happy when they are freely engaging in natural or species-specific behaviours.

Interest in the scientific value of subjective states and dispositions, including happiness, in animals is often traced back to Charles Darwin's (1872/1965) discussion about phylogenetic continuity of emotional displays in *The Expression of the Emotions in Man and Animals*. The book features a striking illustration of a chimpanzee described by Darwin as 'disappointed and sulky'. Earlier, in *The Descent of Man* (1871/1981) Darwin had asserted that 'The lower animals, like man, manifestly feel pleasure and pain, happiness and misery.' He further noted that 'Happiness is never better exhibited than by young animals, when playing together, like our own children.' Later in the text, Darwin says that these conjectures about animals' emotional states were entirely self-evident.

The frequently recounted history of the study of animal behaviour in the aftermath of Darwinian theory (e.g. Boakes, 1984), shows that matters of emotion and personality in animals were deemed neither self-evident nor scientifically promising by Darwin's successors. The experimental, laboratory based, study of animal behaviour in an evolutionary context was

101

largely concentrated on explicitly defined learning, whereas study of behaviours in the natural habitat was probably even more extreme in its dependence upon exact specification of studied behaviours (see Lehner, 1979). Those additional dimensions of mind, emotions and personality, which Darwin found so obvious and fascinating had been almost totally expurgated from animal behaviour study.

Earlier, the Victorian reluctance to measure personality and its individual differences combined an enthusiasm for measuring intellectual endeavours had been decried by Sir Francis Galton, Darwin's brilliant and perpetually inquisitive cousin, who stated the problem in a paper entitled *Measurement of Character* (1884). ' . . . Neither do I see any reason why we should be awed at the thought of examining our fellow creatures as best we may, in respect to other faculties than intellect. On the contrary, I think it anomalous that the art of measuring intellectual faculties should have become highly developed, while that of dealing with other qualities should have been little practiced or even considered.' Although Galton was, of course, referring to humans, 100 years later his argument applies with full force to our hesitancy to measure personality in apes and other non-human primates.

The lexical hypothesis and the big-five

A basis for pursuing scientifically rigorous measurement of personality in non-human primates appeared almost 60 years ago. Allport and Odbert (1936) identified 4505 words from *Websters New International Dictionary* (1925) that were potential personality traits. A personality trait is defined by psychologists as an individual's stable, consistent, and long-lasting mode of adjustment to the environment. Thus, long-lasting traits such as 'irascible' are distinct from short-term states such as 'angry'.

The abundance of personality descriptor terms has two interesting implications. The first is the lexical hypothesis which says that all individual differences in human personality that are salient, and socially relevant will become encoded as a single word, namely a personality descriptor (John, Angleitner & Ostendorf, 1988). The lexical hypothesis is a powerful rationale for using natural language terms to describe personality. The existence of 4505 personality descriptors in English-speaking humans suggests that we have a remarkable ability to discriminate among many different and subtle shades of human personality and temperament.

A capacity for perceptive evaluation of personalities would be expected to emerge in an increasingly talkative, ancestral human in which gossip and general social intelligence about other members of the home clan became strategically and tactically important. Dunbar (1992, 1993) has expanded upon this idea by proposing that, as relative neocortex size increased in emerging humans, language (i.e. gossip) replaced grooming as the principal means for maintaining social cohesion. If the lexical hypothesis and 4505 personality descriptors indicate our ability to apply personality related words to other people (and by generalization to non-human primates), then we probably became perceptive gossip-mongers equipped with an impressive vocabulary of personality descriptors early in the phylogenetic development of language. An obvious adaptive benefit of gossip was that it enabled these emerging humans to learn about the propensities of others without interacting with them directly and possibly having to learn the hard way that they were untrustworthy or malicious.

Scientific, controlled application of our talent for gossip to study of human personality began seriously in the 1930s. One immediate goal was to use factor analysis to organize the thousands of personality descriptors into a few dimensions. Research on factor analysis of personality descriptor ratings over the next 30 years finally led to a consensus that, at the most general level, five basic factors could describe the varied manifestations of human personality (Goldberg, 1982, 1990, 1993; McCrea and Costa, 1987; Norman, 1963; Tupes & Christal, 1961). These five factors are now known as the Big-Five.

Factor I is usually labelled extroversion or surgency. Those at the positive pole are confident, impulsive, and venturesome individuals who actively seek social experiences. Individuals at the opposite pole of Factor I are timid, submissive and lethargic.

Factor II, designated as agreeableness, has been characterized as friendly compliance (Digman, 1989). The positive pole of Factor II includes the descriptors friendly, affectionate, and sympathetic while the negative pole includes irritable, suspicious, self-centred, aggressive and in extreme cases would merge into the psychopathic realm.

Factor III, designated as dependability, is more broadly defined than Factors I and II. Terms such as industrious, well organized, diligent, and self-reliant define the positive pole while fickle, scatter brained, and irresponsible are terms at the negative pole.

Factor IV, designated as emotionality or neuroticism, is related to a susceptibility for experiencing emotional distress, anxiety and chronic negative affect (Digman, 1989; Watson & Clark, 1984). Descriptors such as nervous and unstable define the positive pole of Factor IV while independent and unemotional define the negative pole.

Factor V is now usually characterized as openness. Although it has occasionally been labelled as intellect, Factor V now has a more general meaning than simply intelligence. The factor reflects a readiness to incorporate and explore sources of new experience and information. Thus, creativity and curiosity are descriptors at the positive pole, while dullness and unimaginative are descriptors at the negative pole.

It should be emphasized that many descriptors, averaging about 150 for each factor, have significant factor loadings on the Big-Five (Goldberg, 1990; Norman, 1963). Consequently, a few words cannot fully describe the complex meaning attached to each of the Big-Five. Factors IV (emotionality) and V (openness) have the least number of descriptors in English and Factor II (agreeableness) has the most (Norman, 1963).

Current evidence suggests that each of the Big-Five in humans has a heritability coefficient of about 0.45 (Bouchard, 1994). A biological basis for individual differences in the Big-Five would also be strengthened by a demonstration of the phylogenetic generalizability of the Big-Five to chimpanzees. That issue will now be addressed.

The big-five in chimpanzees

Five- or seven-point rating scales have often been used to rate humans on natural language personality descriptors. Therefore, it might be expected that this approach would have been frequently used to assay non-human primate personality. Since rigorous statistical procedures for assessing reliability and various types of validity have been available for many years (e.g. Wiggins, 1973) fears of that old bugbear anthropomorphism could easily have been assuaged. The lexical approach to personality measurement, however, has been used infrequently by primatologists. Three studies have applied factor analysis to personality descriptors of Old World monkeys (Bolig *et al.*, 1992; McGuire, Raleigh & Pollack, 1994; Stevenson-Hinde & Zunz, 1978). Gold and Maple (1994) also applied factor analysis of personality descriptors to an impressive sample of 298 gorillas from North American zoos and research facilities. All four studies used a principal components factor analysis to identify three major factors. Figueredo, Cox & Rhine (1995) showed that the Stevenson-Hinde factors were also applicable to stumptail macaques through an ingenious application of generalizability theory.

The principal components procedure is commonly used in initial factor-analytic studies (Nunnally, 1978). As with all exploratory factor-analytic techniques, the factor definitions are strongly influenced by the selection of descriptor terms, thus complicating comparison of factors across experiments.

Confirmatory factor analysis is an alternative procedure in which each descriptor is assigned to factors on an *a priori* basis before the analysis is made (Long, 1988). The analysis then determines the factor loading of each descriptor in its pre-assigned factor or factors and the statistical significance of the loadings. Advantages of the confirmatory factor analysis over the exploratory procedures are that descriptor loadings are less influenced by random variation. Furthermore, confirmatory factors have the potential for comparison across experiments, even if the particular descriptors are different.

King and Figueredo (1994*b*, 1997) used confirmatory factor analysis to test the generalizability if the human Big-Five model to zoo chimpanzees. The subjects were 100 chimpanzees at 12 zoological parks in the United States that were participants in the ChimpanZoo project of the Jane Goodall Institute. Ratings were made by 53 observers who were either zoo employees or experienced observers who had previously collected behavioural data for the ChimpanZoo project. Observers used a seven-point scale to rate the chimpanzees on 43 personality descriptors that were distributed across the Big-Five. Overall inter-observer reliability (eta) was 0.753. Confirmatory factor analysis with each descriptor pre-assigned to one of the Big-Five revealed a close fit to the data with no significant amount of variance unaccounted for by the factor model. This was the first evidence that the correlational structure of personality descriptors in chimpanzees is essentially the same as in humans. A subsequent confirmatory factor analysis showed that a revised model consisting of the Big-Five plus a chimpanzee-specific dominance factor more closely fit the data than did the Big-Five model alone. Table 7.1 displays the result of the final factor analysis. Each of the 43 descriptors is shown in its pre-assigned factor along with its factor loading.

Psychological well-being and subjective well-being

Interest in the psychological well-being of captive primates has increased in recent years, partly because of changes in Federal regulations mandating consideration of psychological well-being in captive primates, and partly as a result in a general increase in ethical concerns about captive animals (for reviews, see Novak & Petto, 1991; Novak & Suomi, 1988; Segal, 1989).

Some indicators of psychological well-being have a physiological focus

Table 7.1. *The big-five personality factors in chimpanzees*

Surgery		Agreeableness		Dependability	
Active	0.950[a]	Aggressive	-0.907[ab]	Disorganized	-0.684[a]
Solitary	-0.888[a]	Stingy	-0.862[ab]	Erratic	-0.666[a]
Impulsive	-0.861[a]	Bullying	-0.815[ab]	Decisive	0.579[ab]
Sociable	0.814[a]	Friendly	0.797[a]	Persistent	0.543[ab]
Depressed	-0.776[a]	Sensitive	0.665[a]	Cautious	-0.474[ab]
Timid	-0.704[ab]	Affectionate	0.660[ab]	Reckless	0.459[a]
Dominant	0.326[b]	Irritable	-0.554[a]	Lazy	-0.037
Submissive	-0.309[b]	Helpful	0.550[a]	Predictable	0.033
Playful	0.278	Gentle	0.504[a]		
		Sympathetic	0.475[a]		
		Protective	0.345		
		Manipulative	-0.194[b]		
		Defiant	0.157		
Emotionality		**Openness**		**Chimpanzee dominance**	
Jealous	0.905[a]	Inventive	1.000[a]	Dominant	0.951[a]
Fearful	0.847[ab]	Clumsy	-0.687[a]	Submissive	-0.939[a]
Stable	-0.805[a]	Intelligent	0.531[ab]	Stingy	-0.851[a]
Independent	-0.667[ab]	Inquisitive	0.499[a]	Timid	-0.773[a]
Excitable	0.578[a]	Imitative	0.428[a]	Dependent	-0.544[a]
Dependent	0.425[ab]	Autistic	-0.018	Cautious	-0.495[a]
Unemotional	-0.035			Manipulative	0.238
				Persistent	0.228
				Decisive	0.218
				Intelligent	0.176
				Independent	0.165

[a] Indicates statistically significant factor loading ($p < 0.05$).

including measures of reproductive success, immune system efficiency, and levels of cortisol or testosterone. Other measures focus on prevalence of species-specific behaviours including foraging, low aggression levels, and lack of stereotyped or self-mutilating behaviours. Still other measures involve primates' manipulation of clever devices such as artificial termite mounds or food-dispensing puzzles. Wemelsfelder (1990) proposed an interesting and ingenious approach to definition of psychological well-being by operationally defining boredom, frustration, and depression or helplessness based on animals' responses to novel stimuli.

These different approaches to the psychological well-being of animals are all potentially useful. However, because of the well-known similarity of humans and chimpanzees, a reasonable question is whether there are additional dimensions of chimpanzee mental health not captured by the traditional types of measures described above. Imagine defining human happiness in terms of indices such as lack of self-mutilation or hair plucking, presence of reproductive success or fascination with operating simple mechanical contrivances to earn stale food pellets! The idea is not without risible aspects. Instead, could some measures of human happiness be applied profitably to apes?

Happiness in humans is now subsumed under the more dignified term 'subjective well-being' (Diener, 1984). Happiness did not appear as a term in the *Psychological Abstracts* until 1973, although research on various manifestations of unhappiness has a long history going back to the 19th century origins of psychology. However, philosophers and others worried about the human condition have been trying to analyse the many facets of human happiness for centuries (Tatarkiewicz, 1976). Modern psychological research on happiness has focused on two aspects of subjective well-being: the affective, determined by amount and intensity of positive and negative emotional states and the judgemental, determined by a global evaluation of one's life (Diener & Emmons, 1984; Pavot *et al.*, 1991). Although only the former aspect is potentially applicable to non-human primates, frequencies of positive and negative affect are independent of each other and both are related to global evaluations of subjective well-being (Diener *et al.*, 1985). Furthermore, subjective well-being is more closely related to relative frequencies of positive and negative affect than to intensities of those states (Diener, Sandvik & Pavot, 1991).

The common-sense wisdom that happiness is more affected by temperament and personality than by external circumstances is, in fact, supported by evidence (e.g. Costa & McCrea, 1984). External demographic variables including age, health, income, and marital status are often correlated with subjective well-being, but the correlations are typically small and collectively account for no more than 15% of the total variance in subjective well-being (Campbell, 1976). Subjective well-being is not even related to amount of physical activity (Gauvin, 1989). Life events may have a strong initial effect on subjective well-being, but people eventually adapt to the change and little long-term effect remains (Heady & Wearing, 1989).

Since personality is important in deciding human happiness, the next question is how the Big-Five are specifically related to subjective well-being (SWB). Factor I (surgency) is positively correlated with the amount of time a person experiences positive affect and Factor IV (emotionality) is positively correlated with the amount of time a person experiences negative affect. Similarly, judgemental subjective well-being is positively correlated with Factor I and negatively correlated with Factor IV (Costa & McCrea, 1980; Emmons & Diener, 1985; Hepburn & Eysenck, 1989; McCrea & Costa, 1991). Within the Factor I descriptor complex, sociability is positively correlated with positive affect, and impulsivity is positively correlated with negative affect (Emmons & Diener, 1986).

Factors II (agreeableness) and III (dependability) are positively correlated with positive affect and general life satisfaction (Carp, 1985; McCrea & Costa, 1991). McCrea & Costa noted that this finding is consistent with Freud's widely cited adage that love (Factor II) and work (Factor III) were essential to human happiness.

Factor V (openness) is positively related to both positive and negative affect because of amplifying emotional experiences of both types. It is not related to judgemental subjective well-being (McCrea & Costa, 1991).

Personality and happiness in chimpanzees

We are now closer to being able to answer the question implied by the title to this chapter. However, information about subjective well-being of chimpanzees must first be obtained. The obvious way of obtaining such data is by using the method used for human subjective well-being studies, namely questionnaire ratings. Chimpanzees, unlike humans, cannot rate themselves, but they can be rated by humans with whom they have had a close association. King and Figueredo (1994a) asked zoo observers from the ChimpanZoo project to rate 83 chimpanzees with a seven-point scale on four questions. The first three questions asked about (a) the relative amounts of time the chimpanzees experienced positive and negative feelings, (b) the amount of positive feeling that the chimpanzees received from social interaction with other chimpanzees, and (c) the effectiveness of the chimpanzee in achieving its wishes or goals with its social group. The last question asked the rater to imagine that he or she were to be a chimpanzee for a week. Each chimpanzee was then rated with respect to how desirable this imaginary interspecies transmogrification would be from the rater's perspective.

Interobserver reliabilities (eta) for all four questions exceeded 0.75. Factor analysis confirmed that scores from the four questions could be summed to form a single composite factor score. This composite score will be referred to as SWB. A multiple regression in which SWB was predicted from the Big-Five showed that Factor I (surgency) and IV (emotionality) were the strongest predictors, jointly explaining 36% of the SWB variance with surgency having a positive, and emotionality a negative, weighting. This result is consistent with the previously cited human data showing that these two factors are most strongly associated with subjective well-being. Within Factor I, sociability was positively associated with SWB, again consistent with human data. Impulsivity, however, was not significantly correlated with SWB.

Factor III (dependability) had a significant positive correlation ($r = 0.49$) with SWB. However, Factor II (agreeableness) was absolutely independent of SWB ($r = -0.025$). Thus, although love and work may contribute to the Freudian view of human happiness, we must sadly conclude that chimpanzee happiness is related to work but not love. Factor V (openness) had a significant positive correlation with SWB ($r = 0.34$) in contrast to human data, indicating that this factor mainly affects intensity but not direction of affect. A stepwise multiple regression was performed in which all ten possible pairwise interaction terms involving the Big-Five were entered into the regression equation after each of the Big-Five had been entered individually. None of the interaction terms was significant, a result consistent with human data (McCrea & Costa, 1991). Finally, as with humans (Diener, 1984), neither the age nor the sex of the chimpanzees was related to SWB.

The almost absolute lack of a correlation between agreeableness and SWB in chimpanzees has an interesting methodological implication. A potential criticism of the chimpanzee SWB data is that the raters might have simply liked some chimpanzees more than others and these chimpanzees therefore received the highest SWB scores. Well-liked chimpanzees would be rated high on surgency, with its strong sociability component, agreeableness, dependability and openness, while being rated low on emotionality, since this configuration would be deemed desirable or likeable in chimpanzees just as in humans. The failure of agreeableness to conform to this pattern indicates that the likeability hypothesis of SWB is not correct. Furthermore, a similar hypothesis for human subjective well-being has also not been supported (Pavot et al., 1991).

Conclusions

Data reviewed in this chapter demonstrate that personality and subjective well-being can be reliably measured in chimpanzees and that the relationship between these two types of subjective constructs is remarkably similar in chimpanzees and humans. Nevertheless, the question posed by Iko at the start of this chapter lingers. How do we know if the chimpanzee that seems happy to us really is happy? The answer, of course, is that Iko is ultimately correct and we do not know in any absolute, deductive sense about a chimpanzee's, or a human's subjective well-being. The answer must be indirect, in the form of inferences from observed relationships between subjective well-being measures in chimpanzees and other variables with postulated relationships to that happiness. Confirmation of these postulated relationships contributes to the validity of the constructs. The above evidence gives reason for optimism about the construct validity of chimpanzee subjective well-being.

One might further argue that zoo workers' judgements of personality and subjective well-being in familiar chimpanzees should be more accurate than comparable judgements about people, since chimpanzees do not create all of the situation-specific defences and ploys to disguise inner feelings and dispositions that humans do. The inter-observer reliabilities of personality scores and for subjective well-being scores were actually much greater for the chimpanzee judgements than for humans (McCrea, 1982; McCrea & Costa, 1987; Pavot et al., 1991), a result suggesting that people are better at being chimpanzee psychologists than they are at being human psychologists.

Finally, some may view the contents of this chapter as disappointing, since they do not address anything that can be done to increase the happiness of chimpanzees or other apes. Furthermore, the chapter may seem uncongenial because it emphasizes personality, not external conditions, as a central determinant of happiness.

However, the data described here suggest that anyone interested in the psychological welfare of apes should not be hesitant in assessing personality and subjective well-being using questionnaire items in the same way that these constructs are measured in humans, if well-established procedures for assuring reliability and validity are followed. The hunches, intuitions, and other subjective guesses that people have about an animal's feelings and personalities may be useful scientific indicators of the animal's well-being supplementing measures already used. Use of subjective measures may ultimately lessen the psychological distance that we have placed between ourselves and animals because of not fully applying current techniques for assessing personality and subjective well-being. The consequences should be beneficial to humans and animals alike as well as being scientifically illuminating.

Acknowledgement

This chapter is based on data from the ChimpanZoo project of the Jane Goodall Institute. The project has succeeded during the past six years only through the herculean efforts of the ChimpanZoo director Dr Virginia Landau. I wish to express my deepest gratitude to her.

References

Allport, G. W. & Odbert, H. S. (1936). Trait-names: a psycho-lexical study. *Psychological Monographs*, **47**: No. 211.

Boakes, R. (1984). *From Darwin to Behaviorism*. Cambridge: Cambridge University Press.

Bolig, R. , Price, C. S. , O'Neil, P. L. & Suomi, S. J. (1992). Subjective assessment of reactivity level and personality traits of rhesus monkeys. *International Journal of Primatology*, **13**, 287–306.

Bouchard, T. J. (1994). Genes, environment and personality. *Science*, **264**, 1700–1.

Campbell, A. (1976). Subjective measures of well-being, *American Psychologist*, **31**, 117–24.

Carp, F. M. (1985). Relevance of personality traits to adjustment in group living. *Journal of Gerontology*, **40**, 544–51.

Costa, P. T. Jr. & McCrea, R. R. (1980). Influence of extraversion and neuroticism on subjective well-being: happy and unhappy people. *Journal of Personality and Social Psychology*, **38**, 668–78.

Costa, P. T. Jr. & McCrea, R. R. (1984). Personality as a lifelong determinant of wellbeing. In *Affective Processes in Adult Development and Aging*, ed. C. Z. Malatesta & C. E. Izard, pp. 141–57. Beverly Hills: Sage Publications.

Darwin, C. (1871/1981). *The Descent of Man, and Selection in Relation to Sex*. Princeton: Princeton University Press.

Darwin, C. (1872/1965). *The Expression of the Emotions in Man and Animals*. Chicago: University of Chicago Press.

Diener, E. (1984). Subjective well-being. *Psychological Bulletin*, **95**, 542–75.

Diener, E. & Emmons, R. A. (1984). The independence of positive and negative affect. *Journal of Personality and Social Psychology*, **47**, 1105–17.

Diener, E. , Larson, R. J. , Levine, S. & Emmons, R. A. (1985). Intensity and frequency: dimensions underlying positive and negative affect. *Journal of*

Personality and Social Psychology, **48**, 1253–65.

Diener, E. , Sandvik, E. & Pavot, W. (1991). Happiness is the frequency, not intensity of positive versus negative affect. In *Subjective Well-Being: An Interdisciplinary Perspective*, ed. F. Strack, M. Argyle & N. Schwarz, pp. 119–39. New York: Pergamon Press.

Digman, J. M. (1989). Five robust trait dimensions: development, stability, and utility. *Journal of Personality*, **57**, 195–214.

Dunbar, R. I. M. (1992). Why gossip is good for you. *New Scientist*, 21 November, 28–31.

Dunbar, R. I. M. (1993). Coevolution of neocortical size, group size and language in humans. *The Behavioral and Brain Sciences*, **16**, 681–735.

Emmons, R. A. & Diener, E. (1985). Personality correlates of subjective well-being. *Personality and Social Psychology Bulletin*, **11**, 89–97.

Emmons, R. A. & Diener, E. (1986). Influence of impulsivity and sociability on subjective well-being. *Journal of Personality and Social Psychology*, **50**, 1211–15.

Figueredo, A. J. , Cox, R. L. & Rhine, R. J. (1995). A generalizability analysis of subjective personality assessments in the stumptail macaque and zebra finch. *Multivariate Behavioral Research*, **30**, 167–97.

Galton, F. (1884). Measurement of Character. *Fortnightly Review*, **36**, 179–85.

Gauvin, L. (1989). The relationship between regular physical activity and subjective well-being. *Journal of Sport Behavior*, **12**, 107–14.

Gold, K. & Maple, T. L. (1994). Personality assessment in the gorilla and its utility as a measurement tool. *Zoo Biology*, **13**, 509–22.

Goldberg, L. R. (1982). From ace to zombie: some explorations in the language of personality. In *Advances in Personality Assessment*, vol. 1, ed. C. D. Spielberger & J. N. Butcher, pp. 203–34, Hillsdale: Erlbaum.

Goldberg, L. R. (1990). An alternative 'description of personality': the Big-Five factor structure. *Journal of Personality and Social Structure*, **59**, 1216–29.

Goldberg, L. R. (1993). The structure of phenotypic personality traits. *American Psychologists*, **48**, 26–34.

Heady, B. & Wearing, A. (1989). Personality, life events, and subjective well-being: toward a dynamic equilibrium model. *Journal of Personality and Social Psychology*, **57**, 731–9.

Hepburn, L. & Eysenck, M. W. (1989). Personality, average mood, and mood variability. *Personality and Individual Differences*, **10**, 975–83.

John, O. P. , Angleitner, A. & Ostendorf, F. (1988). The lexical approach to personality: a historical review of trait taxonomic research. *European Journal of Personality*, **2**, 171–203.

King, J. E. & Figueredo, A. J. (1994a). Predicting subjective well-being of zoo chimpanzees from personality traits. Paper presented at XVth Congress of International Primatological Society, Bali, Indonesia, August 1994.

King, J. E. & Figueredo, A. J. (1994b). Personality traits in zoo chimpanzees? Paper presented at the meeting of the Western Psychological Association, Kona, Hawaii, April 1994.

King, J. E. & Figueredo, A. J. (1997). The five-factor model plus dominance in chimpanzee personality. *Journal of Research in Personality*, **31**, 257–71.

Lehner, P. (1979). *Handbook of Ethological Methods*. New York: Garland.

Long, J. S. (1988). *Confirmatory Factor Analysis*. Beverly Hills: Sage Publications.

McCrea, R. R. (1982). Consensual validation of personality traits: evidence from self-reports and ratings. *Journal of Personality and Social Psychology*, **43**, 293–303.

McCrea, R. R. & Costa, P. T. Jr. (1987). Validation of the five-factor model of personality across instruments and observers. *Journal of Personality and Social Psychology*, **66**, 574–83.

McCrea, R. R. & Costa, P. T. Jr. (1991). Adding *Liebe* and *Arbeit*: the full five-factor model and well-being. *Personality and Social Psychology Bulletin*, **17**, 227–32.

McGuire, M. T. , Raleigh, M. J. & Pollack, D. B. (1994). Personality features in velvet monkeys: the effects of sex, age, social status, and group composition. *American Journal of Primatology*, **33**, 1–13.

Norman, W. T. (1963). Toward an adequate taxonomy of personality attributes: replicated factor structure in peer nomination personality ratings. *Journal of Abnormal and Social Psychology*, **66**, 574–83.

Novak, M. A. & Petto, A. J. (ed.) (1991). *Through the Looking Glass: Issues of Psychological Well-Being in Captive Nonhuman Primates.* Washington DC: American Psychological Association.

Novak, M. A. & Suomi, S. J. (1988). Psychological well-being of primates in captivity. *American Psychologist*, **43**, 765–73.

Nunnally, J. C. (1978). *Psychometric Theory.* New York: McGraw-Hill.

Pavot, W. , Diener, E. , Colvin, C. R. & Sandvik, E. (1991). Further validation of the satisfaction with life scale: evidence for the cross-method convergence of the well-being measures. *Journal of Personality Research*, **57**, 149–61.

Segal, E. F. (1989). *Housing, Care and Psychological Well-Being of Captive and Laboratory Primates.* Park Ridge: Noyes Publications.

Stevenson-Hinde, J. & Zunz, M. (1978). Subjective assessment of individual rhesus monkeys. *Primates*, **19**, 473–82.

Tatarkiewicz, W. (1976). *Analysis of Happiness*, The Hague: Martinus Nijhoff.

Tupes, E. C. & Christal, R. E. (1961). *Recurrent Personality Factors Based on Trait Ratings.* ASAF Technical Report 61-97. Lackland Air Force Base: US Air Force.

Watson, D. & Clark, L. A. (1984). Negative affectivity: the disposition to experience aversive emotional states. *Psychological Bulletin*, **96**, 465–90.

Webster's New International Dictionary, 2nd unabridged edn. (1925). Springfield: Merriam.

Wemelsfelder, F. (1990). Boredom and laboratory animal welfare. In *The Experimental Animal in Biomedical Research*, vol. 1, ed. B. E. Rollins & M. L. Kessel, pp. 243–71. Boca Raton: CRC Press.

Wiggins, J. S. (1973). *Personality and Prediction: Principles of Personality Assessment.* Reading: Addison-Wesley.

8

Primate cognition: evidence for the ethical treatment of primates

RICHARD W. BYRNE

Jakob Bronowski, in *The Ascent of Man* (1973), considered it not at all surprising that Galileo – as a man of very great imagination – was prepared to recant scientific beliefs that he knew to be right, at the mere *threat* of torture on the rack: 'his imagination could do the rest'. The implication is that greater ability to imagine personal future events will confer greater scope for mental torment and suffering. (On the same lines, a mountaineer once facetiously remarked that the most important quality for an ice-climber was a complete lack of imagination!) Many people would agree with this diagnosis, when applied to human beings. Yet animal welfare is normally restricted to the demands of the here-and-now. 'Suffering' is viewed as a matter of current deprivation (for instance, of sleep, water, or affiliative social interactions) or current imposition (for instance, of pain, noise, social overload, or other upsetting circumstances). If all non-humans lack the ability to imagine a specific, personal future, and live only in the present, this approach is correct. But can we be sure?

Given the practical difficulties of studying animal imagination, a difficult choice of 'null hypothesis' immediately confronts us. Should we assume, until good evidence is provided otherwise, that only humans can ever suffer in anticipation of imagined future events? This would be consistent with the Judaic–Christian tradition of reserving all higher cognitive function for ourselves, and allow a robust, easily assessed welfare standard to be applied even-handedly to all animals. Or should we assume, until good evidence is provided otherwise, that all species share the human propensity for worrying about the future, and suffering in anticipation? This gives other species the benefit of the doubt, perhaps appropriately, given our admitted uncertainty, and is consistent with the evolutionary continuity argued by evolution.

Neither of these null hypotheses can be rejected on present evidence.

Those people determined – for whatever reason – to hold one or the other will remain undissuaded by the existing data. However, if we instead agree to take a neutral starting point (to take a Bayesian approach, in effect), then I will argue in this chapter that the evidence warrants a tentative division within the animal kingdom; between those species (a majority) in which no evidence of any significant mental anticipation is yet known, and those species (a few) for which there is suggestive evidence of an ability to imagine the future. The implications of this conclusion for welfare of those few species are considerable.

In making this rather controversial distinction, I must make very clear what I mean by anticipation. Animals routinely show behaviour that, in a sense, anticipates the future. Many hibernate, and a very wide range of species show a circadian rhythm with a regular daily period of sleep. It could be said that hibernation 'anticipates' the difficulty of finding food in the temperate winter, and that sleep 'anticipates' the danger of predation or difficulty of feeding at a time when sense organs are not functioning optimally. However, there is every reason to suppose these rhythms are genetically coded, and they probably do not involve imagination of possible, unpleasant personal futures. More borderline is the sense of 'anticipation' in which it might be said that all animal learning anticipates the future. Clearly, a major biological function of learning is to decrease uncertainty about future events and increase an individual's ability to profit from experience. If a jay or a nutcracker stores a cache of nuts, it will later 'anticipate' finding them in a particular place, in the sense that it will look there. If a rat has been shocked in a particular part of an experimental apparatus, it will later 'anticipate' pain in that place, in the sense that it will avoid it. Avoidance learning is rapid, and associated with behaviour and physiological symptoms suggesting that the place where the shock was administered comes to evoke fear. (In the 1930s, the concept of a 'secondary drive', anxiety, was introduced to explain these effects.) What is less clear is whether an animal in such circumstances remembers personal events from its past and imagines these events happening again, as we would in similar circumstances. Certainly, animal learning theory succeeds in explaining the behaviour (and the physiological symptoms) without need for making this supposition. The fact is, it is often difficult to discern a difference in overt behaviour between the memorized consequences of past history ('learning') and the mental construction of a possible personal future ('imagination').

Both have welfare implications. Putting animals into circumstances, associated in their personal history with pain or suffering, may be expected to evoke unpleasant current states and potentially cause suffering. This is already part of the conventional wisdom of animal welfare. However, if any animal has the mental capacity to anticipate aversive events and states that have not yet happened, its capacity to suffer would be much greater. Not constrained by a lack of imagination to its immediately current circumstances, such an animal may brood about what it judges likely to happen to it in the future. It is this sense of anticipation that I shall examine now.

Diagnostics of anticipatory planning

Learning by the mechanisms of associative conditioning has surprising power to mimic the products of reflective thought (e.g. Pfungst, 1911/ 1965). Inevitably, we must rule out conventional explanations of animal learning – classical and operant conditioning – as reasonable candidates before any instance may be considered as possible evidence for anticipatory planning. Equally, we should be aware of the rich possibilities for genetical inheritance to give rise, with little influence of the environment, to behaviours which in ourselves we should assume were planned (e.g. the elaborately co-operative enterprises and extreme 'altruism' of social insects; see Dawkins, 1976). Once more, 'simpler' explanations (simpler in the sense of 'not needing additional theoretical postulates', rather than 'uncomplicated') must be ruled out. The danger of this approach of defining by exclusion is that a motley, heterogeneous category results. Nevertheless, there is no other option open to us at present.

With these provisos, signs of anticipatory planning might potentially be shown in:

- deception, and other manipulative social tactics;
- teaching;
- experiments in which future events need to be predicted for success;
- selection or preparation of tools or other material in advance of need;
- in general, in any instance where past events could not have entirely determined ('conditioned') the appropriate action.

This chapter will mention data from all these varied circumstances, but cannot pretend to be an exhaustive review; the very heterogeneity of the category makes it unlikely that every relevant datum has been covered. Even the list of topics could be extended: at various times, other performances have been held to be indicative of animals having a theory of mind, and hence reflective thoughts. These include mirror self-recognition (Gallup, 1970, 1977, 1982), pretend play (Leslie, 1987; Whiten & Byrne, 1991), and imitation (Bruner, 1972; Whiten & Ham, 1992). There is, of course, debate about whether non-humans do show these behaviours, but that is not the important point here. Rather, the problem is that, supposing they do, it is not clear that that would imply any cognitive abilities relevant to welfare issues. Thus, the ability to react appropriately to self-reflection in a mirror may indicate only the ability to learn how to deal with novel, displaced visual feedback (Heyes, 1994); certainly, the connection between one's reflection and one's thoughts of the future is not obvious. Pretend play in children is divorced in time of acquisition from other indications of theory of mind (Gopnik, 1993): the cognitive underpinnings of pretence in play are not well understood. Also, in practice, it is often difficult to distinguish between the (supposedly innate and cognitively simple) pretence shown by many animals, such as a kitten playing with a ball of wool, and the (supposedly cognitively complex) 'true' pretence of children. Neither does imitation convincingly require reflective thought. It may indicate only the ability to relate visual information to feedback from own actions (Heyes, 1993*b*); in most cases of imitation, there is no need to invoke mental representation of future goals and intentions in order to explain deliberate copying of behaviour patterns (Byrne, 1994). Given these challenges to the *implications* of the behavioural phenomena, I shall not discuss evidence for their occurrence any further in this chapter, concentrating instead on cases where – if alternative explanations can be set aside – performance would indicate some reflective mental ability, having direct relevance to welfare.

Anticipation in deception

Most cases of tactical deception (i.e. tactics that function by deception, but not necessarily by intending the deceit) can be straightforwardly explained as a result of rapid learning by conventional associative mechanisms (Byrne & Whiten, 1992). The fact that most orders of mammal other than primates do not show learned deception is presumably associated with the rapidity of learning needed, because the coincidences of circumstances that allow the tactics to be learned are infrequent in the wild; in the artificial conditions of captivity, pet dogs and cats often acquire tactical deception (Byrne, 1997). However, among a large corpus of primate records of deception, a few records are exceptionally difficult to explain as the products of associative learning or mistaken interpretation (Byrne & Whiten, 1991, 1992); unlike the majority of records, these cases largely come from great apes.

Just two examples will show the diversity of this sort of evidence. Frans Plooij, researching wild chimpanzees at Gombe, describes how he was able to trick an infant chimpanzee who persisted in trying to groom and climb on him (see Byrne & Whiten, 1988), Plooij suddenly acted as if he had seen a distant object of great interest, staring fixedly into the bushes; the infant desisted from her attentions, and set off to investigate. But, when she returned, having found nothing, she 'walked over to me, hit me over the head with her hand and ignored me for the rest of the day'. Only some sort of 'righteous indignation' can reasonably explain the marked change in affect, and this suggests an ability to understand that the deceit was planned. Frans de Waal (1982), working with a captive colony, observed the reaction of a male chimpanzee to an agonistic challenge:

a male, who was sitting with his back to his challenger, showed a grin [indicating fear] on hearing hooting sounds. He quickly used his fingers to push his lips back over his teeth again. The manipulation occurred three times before the grin ceased to appear. After that, the male turned around to bluff back at his rival.

It is difficult to explain such correction of 'leakage' of tell-tale emotion without invoking an ability to anticipate the thought processes of the rival in some way. More recently, Tanner and Byrne (1993) have described a gorilla covering its face with a hand, again apparently aimed at concealing emotional leakage. Each single case like these can be explained away as associative learning (see Byrne & Whiten, 1991, which does so), but the attempts are so cumbersome and implausible that, in the end, anticipatory planning becomes a more parsimonious explanation.

Anticipatory teaching

Like deception, teaching can be defined functionally (Caro & Hauser, 1992): evolved propensities may serve to further learning, with no implication that the 'teacher' anticipates the consequences of its behaviour. However, a few instances seem to imply deliberate, intentional teaching. Boesch (1991) has described the actions of mother chimpanzees whose offspring were not competent at the technically demanding task of nut-cracking using hammer-stone and anvil. One, after her juvenile had placed the nut haphazardly on the anvil, 'took it in her hand, cleaned the anvil, and replaced the piece carefully in the correct position', whereupon the juvenile was successful. After another juvenile had failed in using an irregular-shaped hammer held inefficiently, the mother 'in a very deliberate manner, slowly rotated the hammer into the best position ... it took her a full minute to perform this simple rotation'. The mother then cracked ten nuts, sharing most of the kernels. When her daughter resumed attempts, she was more successful and, crucially, 'always maintained the hammer in the same position as her mother did'. Intentional teaching has also been observed in captivity. Washoe, a chimpanzee taught sign language, adopted and reared an infant chimpanzee; the researchers did not teach the infant sign language, and avoided all signing in her presence (Fouts, Fouts & Van Cantfort, 1989). Nevertheless, the infant acquired at least 51 signs in 5 years, and Washoe was seen to use both demonstration (with careful attention to Loulis' gaze direction) and moulding of Loulis' hands to teach her to sign. In these cases, the intent to teach (rather than simply help or protect) suggests some ability to anticipate the consequences of the youngster's lack of knowledge.

Anticipating future needs and risks

When making tools for food-processing, chimpanzees sometimes fabricate or carefully select the tool in advance and out of sight of the food. Termite-fishing probes may be carried a few hundred metres, after their preparation from the raw plant materials, to the termite mound itself (Goodall, 1986); the rather scarce hammer-stones in West African forests are taken up to a kilometre to sources of nuts (Boesch & Boesch, 1984). In both cases, there is a clear implication that the chimpanzees have specific future purpose in mind when they make their preparations.

Other data, which apparently show anticipation, may in fact not stem from planning or imagination. Chimpanzee termite-fishing is possible for only a limited season each year: when the reproductive, winged alates are preparing to emerge, the mound surface is softer than normal (at points where tubes will be opened to allow emergence) and can be picked open with a fingernail. At the start of this season, chimpanzees begin to travel to and inspect the mounds – which they have previously ignored for months, and which are still often showing no sign of winged alate emergence (Goodall, 1986). Similarly, gorillas show that they anticipate the seasonal and local distribution of plant foods. Mountain gorillas (*G. g. beringei*), as the limited season of bamboo-shoot availability approaches, make excursions down to the bamboo zone, during which they travel rapidly and eat little, returning to normal ranging areas with equal rapidity if no shoots are yet ready to eat (personal observation). Grauer's gorillas (*G. g. graueri*) detour to areas of bamboo, and sometimes go so far as to dig invisible shoots from underground (A. Goodall, 1977). Orangutans and baboons also give the strong impression of going out of their way to check sparse seasonal resources for readiness (A. Russon, R. Barton, personal communications). In all these cases, it may not be that individuals imagine specific future events; instead, they might merely have learnt that food is found in these places, at these times of year. This sense of 'anticipation', although valuable to the animals, has no welfare implications.

More speculative evidence of an ability to imagine the future comes from the considerable risks that chimpanzees sometimes take in their behaviour, risks which can only be justified by very long-term gains. Irregularly, groups of males in several populations make incursions into the range of other communities, sometimes killing single males they encounter (e.g. Goodall *et al.*, 1979). In the end, this 'warfare' has regularly resulted in range extensions by the aggressors, and transfer of females to their group, but these rewards follow only months or years later (Goodall, 1986; Nishida, Takasaki & Takahata, 1990). And, in a single instance, a group of chimpanzee males was seen to kill a leopard cub which they took from the mother, abandoning the body once the cub was dying (Hiraiwa-Hasegawa *et al.*, 1986; Byrne & Byrne, 1988). The risks, of confronting a major predator in the confined space of a narrow cave and attempting to take its offspring, would seem to be great. Possible gains, in terms of either deterrence or population control of leopards, suggest an ability to plan for the uncomfortably distant future. In both these cases, it remains possible to account *post hoc* for the actions as genetically programmed, but the plausibility of doing so can be questioned. The actions raise the distinct possibility of anticipatory planning in the absence of any current need, going far beyond apes' well-documented anticipation in subsistence activities.

Experimental tests of anticipation

The experiment of Premack & Woodruff (1978) which gave rise to the concept 'theory of mind' relied on a chimpanzee's appropriate choice of a solution to another individual's problem. A short film clip depicted the problem, for instance, a shivering man in a bare room with an unlit stove, and a range of photographs were presented, including a box of matches – the 'correct' choice, if the task is perceived by the chimpanzee as one of anticipating the portrayed individual's hopes for the future. The experiment is problematic, not least because it was conducted on only one animal, and it is difficult to exclude other explanations – in the instance mentioned, for instance, the 'correct' choice might reflect only an association between stove and matches (see Premack, 1988).

Theory of mind in primates has been tested also by giving an animal a choice of informant, between a person who has relevant knowledge and one who can only guess (Premack, 1988; Povinelli, Nelson & Boysen, 1990; Povinelli, Parks & Novak, 1991). For instance, one of two boxes is baited so that the primate cannot – but a human can – see which it is. This person and another – one who was either out of the room at the time, covered their head, or stood behind a screen – then offer conflicting advice to the animal. Whose advice does it follow? The experiments show that chimpanzees, but not rhesus monkeys, reliably follow the 'knowing' advisor, although not on the first trial. However, it is again difficult to exclude alternative explanations, such as chimpanzee preference for individuals who have been most closely associated with the apparatus (see Heyes, 1993*a*).

An ability to comprehend intentions is perhaps most straightforwardly implied by distinguishing correctly between accident and deliberate malice. This has also been shown experimentally in a chimpanzee, who was deprived of a valued drink, in two circumstances: on one occasion, the loss was through an 'obvious accident' by an apparently well-intentioned human, who tripped; on another, a different person brought the drink but then deliberately poured it on the ground in front of the chimpanzee (Povinelli, 1991). In future, the chimpanzee consistently chose the 'clumsy' person for its drink requests. As with Premack and Woodruff's (1978) seminal experiment, this relies on a single chimpanzee. Furthermore, the chimpanzee may have been influenced by the demeanour of the human who deprived it of juice. If the 'mean' human acts naturally, as they did in this experiment, then they will be conveying a threatening image. But, if they were to adopt a bland exterior, then their actions and demeanour would be strangely at variance, confusing even to a human observer. There may be no way to avoid this contamination of possible causes with experimental design.

Understanding the goals and intentions of others can allow prediction of their future needs, to mutual gain: for instance, in a two-person collaborative task where one can see what to do and the other is able to do it. Simple trial-and-error learning can also lead to efficient co-operation in such a task; but chimpanzees, having learned one role, can immediately switch to the other, whereas rhesus monkeys cannot and must learn the new role from scratch (Povinelli, Nelson & Boysen, 1992; Povinelli, Parks & Novak, 1992). This evidence suggests that chimpanzees understand the task, in such a way that they can anticipate the needs of the partner, whereas monkeys may simply learn 'rules of thumb' that work without any insight as to why. Once again, this is consistent with an ability to work out future possibilities, in apes but not monkeys. As so often in laboratory work, individuals of just one species of monkey were used (the rhesus, *Macaca mulatta*). Generalizing their failure to all 29 genera of monkeys is evidently risky, and in particular it should be remembered that baboons are much larger-brained than macaques. However, field data on tactical deception, in which baboons are heavily represented, similarly gives no evidence of any ability to understand intentions of others.

The most compelling experimental evidence of theory of mind in primates comes from an experiment originally carried out with infant macaque monkeys and their mothers (Cheney & Seyfarth, 1990). The mothers were shown a predator, either in the presence of their infant or not. The monkeys did not attempt to alert ignorant offspring more than knowledgeable ones, consistent with the idea that monkeys lack the ability to attribute mental states. However, when Boysen (1997) replicated this experiment with chimpanzees, they did show differentiation between informed and ignorant companions. In this case, the 'predator' was a human with a dart gun, and instead of mothers and infants Boysen used highly affiliated pairs of young chimpanzees. When both chimpanzees could see the human, or the human lacked a dart gun, or the observing chimpanzee could see that its friend was not free to move into the 'danger zone', the reaction was negligible. But, when the view of the human was restricted to one chimpanzee and the other chimpanzee was free to come out, then the observer displayed and called loudly. These displays were effective, since the other chimpanzee fear-grinned and refused to come out under these circumstances.

Implications for welfare

In almost every case, the evidence – though not always conclusive – has pointed to one group of primates, the great apes, as having some ability to imagine future possible states of affairs. The well-studied chimpanzee has given the most convincing signs, but this may change as the data improves: already, all four species (gorilla, orangutan and pygmy chimpanzee as well) have shown intentional deception. If we accept this converging evidence of anticipatory planning, what are its implications?

Animals which can – in however rudimentary a way – predict future possibilities, instead of merely reacting on the basis of past reinforcement histories without envisaging possible consequences, are mentally very much closer to humans than those that cannot. To assess their welfare needs properly, the safest course would be to treat them as humans. This would not be anthropomorphism, but a consequence of a rational assessment of the evidence. Instead of being concerned only with what happens to chimpanzees, we should be concerned with what they might think will happen. An incident described by the team who studied the sign language of Washoe and her adopted infant, mentioned above, illustrates the potential problem. Before the adoption, Washoe in fact gave birth herself, but the baby died and the mother was naturally upset. The researchers, having managed to discover a new-born chimpanzee which they could acquire for her to adopt, used sign language to tell Washoe. She became very animated again, but when the infant arrived, 'her high excitement evaporated' and at first she rejected it completely (Fouts *et al.*, 1989; fortunately she did later adopt the infant). It would seem that the researchers had unintentionally conveyed to Washoe that her own baby would return, giving rise to the disappointment when her anticipation was unfulfilled. Many researchers who carry out non-invasive experiments with chimpanzees probably take such possibilities into account already; and probably they do not do so with their dogs and cats, however much they are fond of them. However, I believe that current evidence proscribes any further invasive experimentation on great apes, beyond what could be ethically carried out on humans who are denied or incapable of giving informed consent.

It is much harder to use the data of this chapter to justify a further step, as advocated by the signatories to 'The great ape project' (Cavalieri & Singer, 1993), of giving great apes human rights. With such 'rights' unavailable to so many humans, it is unclear what good this gesture would do. Nor would awarding human rights to apes have any obvious beneficial implications for the conservation of the wild populations, or for the many individuals already in zoos, who would certainly not benefit from release.

References

Boesch, C. (1991). Teaching in wild chimpanzees. *Animal Behaviour*, **41**, 530–32.
Boesch, C. & Boesch, H. (1984). Mental map in wild chimpanzees: an analysis of hammer transports for nut cracking. *Primates*, **25**, 160–70.
Boysen, S. (1997). A picture is worth a thousand M&Ms: representational flexibility in chimpanzees. *American Journal of Primatology*, **42**, 86.
Bronowski, J. (1973). *The Ascent of Man*. London: Science Horizons Inc.
Bruner, J. S. (1972). Nature and uses of immaturity. *American Psychologist*, **27**, 687–708.
Byrne, R. W. (1994). The evolution of intelligence. In *Behaviour and Evolution*, ed. P. J. B. Slater & T. R. Halliday, pp. 223–65. Cambridge: Cambridge University Press.
Byrne, R. W. (1997). What's the use of anecdotes? Attempts to distinguish

psychological mechanisms in primate tactical deception. In *Anthropomorphism, Anecdotes, and Animals: The Emperor's New Clothes?*, ed. R. W. Mitchell, N. S. Thompson & L. Miles, pp. 134–50. New York: State University of New York Press.

Byrne, R. W. & Byrne, J. M. (1988). Leopard killers of Mahale. *Natural History*, **97**, 22–6.

Byrne, R. W. & Whiten, A. (1988). Towards the next generation in data quality: a new survey of primate tactical deception. *Behavioral and Brain Sciences*, **11**, 267–73.

Byrne, R. W. & Whiten, A. (1991). Computation and mindreading in primate tactical deception. In *Natural Theories of Mind*, ed. A. Whiten, pp. 127–41. Oxford: Basil Blackwell.

Byrne, R. W. & Whiten, A. (1992). Cognitive evolution in primates: evidence from tactical deception. *Man*, **27**, 609–27.

Caro, T. M. & Hauser, M. D. (1992). Is there teaching in non-human animals? *Quarterly Review of Biology*, **67**, 151–74.

Cavalieri, P. & Singer, P. (1993). *The Great Ape Project: Equality Beyond Humanity*. London: Forth Estate.

Cheney, D. & Seyfarth, R. (1990). Attending to behaviour versus attending to knowledge: examining monkeys' attribution of mental states. *Animal Behaviour*, **40**, 742–53.

Dawkins, R. (1976). *The Selfish Gene*. Oxford: Oxford University Press.

Fouts, R. S. , Fouts, D. H. & Van Cantfort, T. E. (1989). The infant Loulis learns signs from cross-fostered chimpanzees. In *Teaching Sign Language to Chimpanzees*, ed. R. A. Gardner, B. T. Gardner & T. E. Van Cantfort. New York: State University of New York Press.

Gallup, G. G. (1970). Chimpanzees: self-recognition, *Science*, **167**, 86–7.

Gallup, G. G. (1977). Self-recognition in primates. *American Psychologist*, **32**, 329–38.

Gallup, G. G. (1982). Self-awareness and the emergence of mind in primates. *American Journal of Primatology*, **2**, 237–48.

Goodall, A. G. (1977). Feeding and ranging behaviour of a mountain gorilla group (*Gorilla gorilla beringei*) in the Tshibinda–Kahuzi region (Zaire). In *Primate Ecology*, ed. T. H. Clutton-Brock, pp. 450–79. New York: Academic Press.

Goodall, J. (1986). *The Chimpanzees of Gombe: Patterns of Behaviour*. Cambridge, MA: Harvard University Press.

Goodall, J. , Bandora, A. , Bergmann, E. , Busse, C. , Matama, H. , Mpongo, E. , Pierce, A. & Riss, D. (1979). Intercommunity interactions in the chimpanzee population of the Gombe National Park. In *The Great Apes*, ed. D. Hamburg & E. R. McCown, pp. 13–54. Menlo Park: Benjamin Cummings.

Gopnik, A. (1993). How we know our minds: the illusion of first-person knowledge of intentionality. *Behavioural and Brain Sciences*, **16**, 1–14.

Heyes, C. M. (1993a). Anecdotes, training, trapping and triangulating. *Animal Behaviour*, **46**, 177–88.

Heyes, C. M. (1993b). Imitation, culture and cognition. *Animal Behaviour*, **46**, 999–1010.

Heyes, C. M. (1994). Reflections on self-recognition in primates. *Animal Behaviour*, **47**, 909–19.

Hiraiwa-Hasegawa, M. , Byrne, R. W. & Takasaki, H. & Byrne, J. M. (1986). Aggression towards large carnivores by wild chimpanzees of Mahale

Mountains National Park, Tanzania. *Folia Primatologica*, **47**, 8–13.

Leslie, A. M. (1987). Pretense and representation: the origins of 'theory of mind'. *Psychological Review*, **94**, 412–26.

Nishida, T. , Takasaki, H. & Takahata, Y. (1990). Demography and reproductive profiles. In *The Chimpanzees of the Mahale Mountains*, ed. T. Nishida, pp. 63–97. Tokyo: University of Tokyo Press.

Pfungst, O. (1911/1965). *Clever Hans, the Horse of Mr von Osten*. New York: Holt, Rinehart & Winston.

Povinelli, D. J. (1991). Social intelligence in monkeys and apes. PhD Thesis, Yale University, New Haven, Connecticut.

Povinelli, D. J. , Nelson, K. E. & Boysen, S. T. (1990). Inferences about guessing and knowing in chimpanzees (*Pan troglodytes*). *Journal of Comparative Psychology*, **104**, 203–10.

Povinelli, D. J. , Nelson, K. E. & Boysen, S. T. (1992). Comprehension of role reversal in chimpanzees: evidence of empathy? *Animal Behaviour*, **43**, 633–40.

Povinelli, D. J. , Parks, K. A. & Novak, M. A. (1991). Do rhesus monkeys (*Macaca mulatta*) attribute knowledge and ignorance to others? *Journal of Comparative Psychology*, **105**, 318–25.

Povinelli, D. J. , Parks, K. A. & Novak, M. A. (1992). Role reversal by rhesus monkeys but no evidence of empathy. *Animal Behaviour*, **43**, 269–81.

Premack, D. (1988). 'Does the chimpanzee have a theory of mind?' revisited. In *Machiavellian Intelligence: Social Expertise and the Evolution of Intellect in Monkeys, Apes and Humans*, ed. R. W. Byrne & A. Whiten, pp. 94–110. Oxford: Clarendon Press.

Premack, D. & Woodruff, G. (1978). Does the chimpanzee have a theory of mind? *The Behavioral and Brain Sciences*, **4**, 515–26.

Tanner, J. E. & Byrne, R. W. (1993). Concealing facial evidence of mood: evidence for perspective-taking in a captive gorilla? *Primates*, **34**, 451–6.

de Waal, F. B. M. (1982). *Chimpanzee Politics*. London: Jonathan Cape.

Whiten, A. & Byrne, R. W. (1991). The emergence of meta-representation in human ontogeny and primate phylogeny. In *Natural Theories of Mind*, ed. A. Whiten, pp. 267–81. Oxford: Basil Blackwell.

Whiten, A. & Ham, R. (1992). On the nature and evolution of imitation in the animal kingdom: reappraisal of a century of research. In *Advances in the Study of Behavior*, Vol. 21, ed. P. J. B. Slater, J. S. Rosenblatt, C. Beer & M. Milinski, pp. 239–83. New York: Academic Press.

Part III

Animal welfare

9

Animal welfare: the concept of the issues

DONALD M. BROOM

Introduction

The first part of this chapter concerns the concept of animal welfare. A definition of welfare is presented and then discussed in relation to other relevant concepts such as stress and animals' needs and feelings. In the final section of this chapter, there is a discussion of the circumstances in which welfare can be poor and where this is perceived by the general public to pose ethical problems.

Requirements for a definition of animal welfare

Welfare is a term which is restricted to animals including man. It is regarded as particularly important by many people but requires strict definition if it is to be used effectively and consistently. A clearly defined concept of welfare is needed for use in precise scientific measurements, in legal documents and in public statements or discussion. If animal welfare is to be compared in different situations or evaluated in a specific situation, it must be assessed in an objective way. The assessment of welfare should be quite separate from any ethical judgement but, once an assessment is completed, it should provide information which can be used to take decisions about the ethics of a situation.

An essential criterion for a useful definition of animal welfare is that it must refer to a characteristic of the individual animal rather than something given to the animal by man. The welfare of an individual may well improve as a result of something given to it, but the thing given is not itself welfare. The loose use of welfare with reference to payments to poor people is irrelevant to the scientific or legal meaning. However, it is accurate to refer to changes in the welfare of an initially hungry person who uses a payment to obtain food and then eats the food. We can use the word welfare in relation to a person, as above, or an animal which is wild or is captive on a farm, in a zoo, in a laboratory, or in a human home. Effects on welfare which can be described include those of disease, injury, starvation, beneficial stimulation, social interactions, housing conditions, deliberate ill treatment, human handling, transport, laboratory procedures, various mutilations, veterinary treatment or genetic change by conventional breeding or genetic engineering.

We have to define welfare in such a way that it can be readily related to other concepts such as: needs, freedoms, happiness, coping, control, predictability, feelings, suffering, pain, anxiety, fear, boredom, stress and health.

Welfare definition

If, at some particular time, an individual has no problems to deal with, that individual is likely to be in a good state including good feelings and indicated by body physiology, brain state and behaviour. Another individual may face problems in life which are such that it is unable to cope with them. Coping implies having control of mental and bodily stability and prolonged failure to cope results in failure to grow, failure to reproduce or death. A third individual might face problems but, using its array of coping mechanisms, be able to cope but only with difficulty. The second and third individuals are likely to show some direct signs of their potential failure to cope or difficulty in coping and they are also likely to have had bad feelings associated with their situations. The welfare of an individual is its state as regards its attempts to cope with its environment (Broom, 1986). This definition refers to a characteristic of the individual at the time. The origin of the concept is how well the individual is faring or travelling through life and the definition refers to state at a particular time (for further discussion, see Broom, 1991a, 1993; Broom & Johnson, 1993). The concept refers to the state of the individual on a scale from very good to very poor. This is a measurable state and any measurement should be independent of ethical considerations. When considering how to assess the welfare of an individual, it is necessary to start with knowledge of the biology of the animal. The state may be good or poor, however, in either case, in addition to direct measures of the state, attempts should be made to measure those feelings which are a part of the state of the individual.

The assessment of welfare is summarized in Table 9.1 and indicators of good and poor welfare are listed in Table 9.2 and Table 9.3. Most indicators will help to pinpoint the state of the animal wherever it is on the scale from very good to very poor. Some measures are most relevant to

Table 9.1. *Summary of welfare assessment*

General methods	Assessment
Direct indicators of poor welfare	How poor
Tests of (a) avoidance and (b) positive preference	Extent to which: (a) animals have to live with avoided situations or stimuli (b) what is strongly preferred is available
Measure of ability to carry out normal behaviour and other biological functions.	How much important normal behaviour or physiological or anatomical development cannot occur

Table 9.2. *Measures of poor welfare*

Reduced life expectancy
Reduced ability to grow or breed
Body damage
Disease
Immunosuppression
Physiological attempts to cope
Behavioural attempts to cope
Behaviour pathology
Self-narcotization
Extent of behavioural aversion shown
Extent of suppression of normal behaviour
Extent to which normal physiological processes and anatomical development are
prevented

(From Broom & Johnson, 1993.)

Table 9.3. *Measures of good welfare*

Variety of normal behaviours shown
Extent to which strongly preferred behaviours can be shown
Physiological indicators of pleasure
Behavioural indicators of pleasure

(From Broom & Johnson, 1993.)

short-term problems, such as those associated with human handling or a brief period of adverse physical conditions, whereas others are more appropriate to long-term problems. (For a detailed discussion of measures of welfare, see Broom, 1988; Fraser & Broom, 1990; and Broom & Johnson, 1993.)

Some signs of poor welfare arise from physiological measurements. For instance, increased heart-rate, adrenal activity, adrenal activity following

ACTH challenge, or reduced immunological response following a challenge, can all indicate that welfare is poorer than in individuals which do not show such changes. Care must be taken when interpreting such results, as with many other measures described here. The impaired immune system function and some of the physiological changes can indicate what has been termed a pre-pathological state (Moberg, 1985).

Behavioural measures are also of particular value in welfare assessment. The fact that an animal avoids an object or event strongly gives information about its feelings and hence about its welfare. The stronger the avoidance, the worse the welfare whilst the object is present or the event is occurring. An individual which is completely unable to adopt a preferred lying posture despite repeated attempts will be assessed as having poorer welfare than one which can adopt the preferred posture. Other abnormal behaviour such as stereotypies, self-mutilation, tail-biting in pigs, feather-pecking in hens or excessively aggressive behaviour indicates that the perpetrator's welfare is poor.

In some of these physiological and behavioural measures it is clear that the individual is trying to cope with adversity and the extent of the attempts to cope can be measured. In other cases, however, some responses are solely pathological and the individual is failing to cope. In either case, the measure indicates poor welfare.

Disease, injury, movement difficulties and growth abnormality all indicate poor welfare. If two housing systems are compared in a carefully controlled experiment and the incidence of any of the above is significantly increased in one of them, the welfare of the animals is worse in that system. The welfare of any diseased animal is worse than that of an animal which is not diseased but much remains to be discovered about the magnitude of the effects of disease on welfare. Little is known about how much suffering is associated with different diseases. A specific example of an effect on housing conditions which leads to poor welfare is the consequence of severely reduced exercise for bone strength. In studies of hens (Knowles & Broom, 1990; Norgaard-Nielsen, 1990) those birds which could not sufficiently exercise their wings and legs because they were housed in battery cages had considerably weaker bones than those birds in percheries which could exercise. Similarly, Marchant & Broom (1994, 1996) found that sows in stalls had leg bones only 65% as strong as sows in group-housing systems. The actual weakness of bones means that the animals are coping less well with their environment, so welfare is poorer in the confined housing. If such an animal's bones are broken, there will be considerable pain and the welfare will be worse. Pain may be assessed by aversion, physiological measures, the effects of analgesics, e.g. Duncan *et al.* (1991) or by the existence of neuromas (Gentle, 1986). Whatever the measurement, data collected in studies of animal welfare gives information about the position of the animal on a scale of welfare from very good to very poor.

The majority of indicators of good welfare which we can use are obtained by studies demonstrating positive preferences by animals. Early studies of this kind included that by Hughes and Black (1973) showing that hens given a choice of different kinds of floor to stand on did not choose what biologists had expected them to choose. As techniques of preference tests developed, it became apparent that good measures of strength of preference were needed. Taking advantage of the fact that gilts preferred to lie in a pen adjacent to other gilts, van Rooijen (1980) offered them the choice of different kinds of floors which were either in pens next to another gilt or in pens further away. With the floor preference titrated against the social preference, he was able to get better information about strength of preference. A further example of preference tests, in which operant conditioning with different fixed ratios of reinforcement were used, is the work of Arey (1992). Pre-parturient sows would press a panel for access to a room containing straw or one containing food. Up to two days before parturition they pressed, at ratios of 50–300 per reinforcement, much more often for access to food than for access to straw. At this time, food was more important to the sow than straw for manipulation or nest-building. However, on the day before parturition, at which time a nest would normally be built, sows pressed just as often, at fixed ratio 50–300, for straw as for food. Another indicator of the effort which an individual is willing to use to obtain a resource is the weight of door which is lifted. Manser *et al.* (1996), studying floor preferences of laboratory rats, found that rats would lift a heavier door to reach a solid floor on which they could rest than to reach a grid floor.

The third general method of welfare assessment listed in Table 9.1 involves measuring what behaviour and other functions cannot be carried out in particular living conditions. Hens prefer to flap their wings at intervals but cannot in a battery cage whilst veal calves and some caged laboratory animals try hard to groom themselves thoroughly but cannot in a small crate, cage or restraining apparatus.

In all welfare assessment, it is necessary to take account of individual variation in attempts to cope with adversity and in the effects which adversity has on the animal. When pigs have been confined in stalls or tethers for some time, a proportion of individuals show high levels of stereotypies whilst others are very inactive and unresponsive (Broom, 1987). There may also be a change with time spent in the condition in the amount and type of abnormal behaviour shown (Cronin & Wiepkema, 1984). In rats, mice and tree shrews it is known that different physiological and behavioural responses are shown by an individual confined with an aggressor and these responses have been categorized as active and passive coping (von Holst, 1986; Koolhaas, Schuurmann & Fokema, 1983; Benus, 1988). Active animals fight vigorously whereas passive animals submit. A study of the strategies adopted by gilts in a competitive social situation showed that some sows were aggressive and successful, a second category of animals defended vigorously if attacked whilst a third category of sows avoided social confrontation if possible. These categories of animals differed in their adrenal responses and in reproductive success (Mendl, Zanella & Broom, 1992). As a result of differences in the extent of different physiological and behavioural responses to problems, it is necessary that any assessment of welfare should include a wide range of measures. Our knowledge of how the various measurements combine to indicate the severity of the problem must also be improved.

Welfare: deduced from measurements and varying over a range

If welfare were viewed as an absolute state which either existed or did not exist, then the concept of welfare would be of little use when discussing the effects on individuals of various conditions in life or of potentially harmful or beneficial procedures. It is essential that the concept be defined in such a way that welfare is amenable to measurement. Once the possibility of measurement is accepted, welfare has to vary over a range. If there is a scale of welfare and the welfare of an individual might improve on this scale, it must also be possible for it to go down the scale. There are many scientists assessing the welfare of animals who accept that welfare can get better or can get poorer. It is therefore illogical to try to use welfare as an absolute state or to limit the term to the good end of the scale. Welfare can be poor as well as good.

The view of welfare as referring only to something good or conducive to a good or preferable life (Tannenbaum, 1991) is not tenable if the concept is to be practically and scientifically useful. Fraser (1993), referring to well-being as the state of the animal, advocates assessing it in terms of level of biological functioning such as injury or malnutrition, extent of suffering and amount of positive experience. However, despite using well-being to refer to scales of how good the animal's condition is, some of his statements explaining well-being imply only a good state of the animal, a limitation which is neither logical nor desirable. Fraser (1993) does, however, follow Broom (1986) and Broom and Johnson (1993) in drawing a conceptual parallel with the term 'health', which is encompassed within the term welfare. Like welfare, health can refer to a range of states and can be qualified as either 'good' or 'poor'. Health can also imply absence of illness or injury. Welfare has the same range of colloquial meaning but when the term is used precisely, it must mean the range of states and it must be possible to refer to poor welfare and good welfare.

Welfare and needs

Animals have a range of functional systems controlling body temperature, nutritional state, social interactions, etc. (Broom, 1981). Together, these functional systems allow the individual to control its interactions with its environment and hence to keep each aspect of its state within a tolerable range. The allocation of time and resources to different physiological or behavioural activities, either within a functional system or between systems, is controlled by motivational mechanisms. When an animal is actually or potentially homeostatically maladjusted, or when it must carry out an action because of some environmental situation, we say that it has a need. A need can therefore be defined as a requirement, which is fundamental in the biology of an animal, to obtain a particular resource or respond to a particular environmental or bodily stimulus (Broom & Johnson, 1993). When needs are not satisfied, welfare will be poorer than when they are satisfied. The degree to which welfare is poor will vary and this has to be scientifically assessed.

Some needs are for particular resources, such as water or heat, but control systems have evolved in animals in such a way that the means of obtaining a particular objective has become important to the individual animal. The animal may need to perform a certain behaviour and may be seriously affected if unable to carry out the activity, even in the presence of the ultimate objective of the activity, for example rats and ostriches will work, in the sense of carrying out actions which result in food presentation, even in the presence of food. In the same way, pigs need to root in soil or some similar substratum (Hutson, 1989), hens need to dust-bathe (Vestergaard, 1980) and both of these species need to build a nest before giving birth or laying eggs (Brantas, 1980; Arey, 1992). In all of these different examples, the need itself is not physiological or behavioural, but may be satisfied only when some physiological imbalance is prevented or rectified, or when some particular behaviour is shown.

Some needs are associated with feelings, which might also be called subjective experiences, and these feelings are likely to change when the need is satisfied. If the existence of a feeling increases the chances that the individual will carry out some adaptive action and hence be more likely to survive, the capacity to have such a feeling is likely to have evolved by natural selection. Furthermore, if the state of an individual in certain conditions is desirable from an evolutionary viewpoint, there should be a propensity for that individual to have good feelings. On the other hand, if a state is one which should be quickly altered, it should be associated with unpleasant feelings which prompt avoidance or some other action. Feelings are part of a mechanism to achieve an end, just as adrenal responses or temperature regulatory behaviour are mechanisms to achieve an end.

When there are no needs which have to be satisfied immediately and the animal's welfare is good, the animal is likely to experience positive feelings. Likewise, when there are unsatisfied needs and welfare is poor, there will often be bad feelings. Feelings will usually result in changed preferences, hence preferences can give some useful information about needs. Other information about needs is obtained by observing the abnormalities of behaviour and physiology which result when needs are not satisfied.

Needs vary in urgency and the consequences if they are not satisfied range from those which are life-threatening to those which are relatively harmless in the short term (see Broom & Johnson, 1993). This range of meaning of need can be expressed in German by two words *Bedarf* and *Bedürfniss*. A *Bedarf* is a need which must be satisfied if life is to continue whereas a *Bedürfniss* is a need which the individual wishes to be satisfied. Since we know that strong preferences by an individual for or against a resource or activity usually relate to something important for the biological success of that individual, a *Bedürfniss* has to be considered very carefully in relation to welfare.

Welfare and feelings

The subjective feelings of an animal are an extremely important part of its welfare (Broom, 1991*b*). Suffering is a negative unpleasant subjective feeling which should be recognized and prevented wherever possible. However, whilst we have many measures which give us some information about injury, disease and both behavioural and physiological attempts to cope with the individual's environment, fewer studies tell us about the feelings of the animal. Information can be obtained about feelings using preference studies but this must be complemented by the other information about welfare mentioned above.

As discussed above, feelings are aspects of an individual's biology which must have evolved to help in survival (Broom, 1998), just as aspects of anatomy, physiology and behaviour have evolved, so it is not logical to concentrate on feelings to the exclusion of other mechanisms when defining welfare. It is also possible, as with any other aspect of the biology of an individual, that some feelings do not confer any advantage on the animal but are epiphenomena of neural activity (Broom & Johnson, 1993). If the definition of welfare were limited to the feelings of the individual, as has been proposed by Duncan and Petherick (1991), it would not be possible to refer to the welfare of a person or an individual of another species which had no feelings because it was asleep, or anaesthetized, or drugged, or suffering from a disease which affects awareness. A further problem, if only feelings were considered, is that a great deal of evidence about welfare like the presence of neuromas, extreme physiological responses or various abnormalities of behaviour, immunosuppression, disease, inability to grow and reproduce, or reduced life expectancy would not be taken as evidence of poor welfare unless bad feelings could be demonstrated to be associated with them. Evidence about feelings must be considered, for it is important in welfare assessment, but to neglect so many other measures is illogical and harmful to the assessment of welfare, and hence to attempts to improve welfare.

In some areas of animal welfare research, it is difficult to identify the subjective experiences of an animal experimentally. For example, it would be difficult to assess the effects of different stunning procedures using preference tests. Disease effects are also difficult to assess using preference tests. There are also problems in interpreting strong preferences for harmful foods or drugs. However, research on the best housing conditions and handling procedures for animals can benefit greatly from studies of preferences which give information about the subjective experiences of animals. Both preference studies and direct monitoring of welfare have an important role in animal welfare research.

Welfare and stress

The word stress should be used for that part of poor welfare which involves failure to cope. If the control systems regulating body state and responding to dangers are not able to prevent displacement of state outside the tolerable range, a situation of different biological importance is reached. The use of the term stress should be restricted to the common public use of the word to refer to a deleterious effect on an individual (see Broom &

Johnson, 1993 for more detailed information on this subject). A definition of stress as just a stimulation or an event which elicits adrenal cortex activity is of no scientific or practical value. A precise criterion for what is adverse for an animal is difficult to find but one indicator is whether there is, or is likely to be, an effect on biological fitness. Stress can be defined as an environmental effect on an individual which over-taxes its control systems and reduces its fitness or seems likely to do so (Broom & Johnson, 1993; see also Broom, 1983; Fraser & Broom, 1990). Using this definition, the relationship between stress and welfare is very clear. First, whilst welfare refers to a range in the state of the animal from very good to very poor, whenever there is stress, welfare is poor. Secondly, stress refers only to situations where there is failure to cope but poor welfare refers to the state of the animal both when there is failure to cope and when the individual is having difficulty in coping. It is very important that this latter kind of poor welfare is included in the definition of welfare as well as the occasions when an animal is stressed. For instance, if a person is severely depressed or if an individual has a debilitating disease, but there is complete recovery with no long-term effects on fitness, then it would still be appropriate to say that the welfare of the individuals was poor at the time of the depression or disease.

Welfare assessment and ethics

Scientific measurement can be used in an investigation of welfare, for example, in a comparison of different animal transport procedures. However, a pertinent question in such studies is to what extent can, or should, the steps in the investigation depend upon ethical considerations? Tannenbaum (1991) argues that welfare is a concept in which values are inextricably involved, so it is not possible to separate what does and does not involve ethics. This is a confusing use of the term values and fails to provide a basis for welfare investigation.

There are four components in a study like that of methods of transport of farm animals. The first is to decide that there is a problem. Ethical considerations are involved here. For instance, it is considered that welfare of farm animals should not be very poor and that welfare may be worse during one transport method than during another. The second component in the study may be to make a scientific comparison of the welfare of the animals during these methods of transport. In this step, the only ethical consideration is that the scientist should be as objective as possible in selecting measurements and in carrying out the study. The scientist should take care to use all possible information about the biology of the animal and the likely environmental effects on the animal when selecting measures. The third component, that is when the measurements are made and analysed, like the second, must be objective and independent of any ethical view about which method of transport is desirable. When the scientific process is completed and the results are presented, ethical decisions can be taken. This is the fourth component. Ethical values are involved in the first and fourth components of the process but only scientific values should be involved in components two and three.

Actual and publicly perceived welfare problem areas

Members of the general public are usually most disturbed by reports of pain or disturbing and bizarre images which concern animals with which they can readily identify. The injured or emaciated dog or horse elicits a greater response from the average person than a similarly injured or emaciated rat, sheep or chicken. The term welfare refers to all animals even if there is variation in the sophistication of the life control mechanisms and hence in the variety of ways in which welfare can be poor.

The nature of human use of an animal or interaction with it has no effect on the extent to which the animal can suffer or be otherwise adversely affected (Broom, 1989). There is an illogical tendency for people to be more concerned about pets than about animals kept in large numbers or largely hidden from their view. Suppose that a rabbit has a certain level of injury or disease. Its welfare is just as poor whether it is a pet, a laboratory animal, a farmed animal or a wild animal.

The most important influences on the welfare of most animals are the living conditions during the majority of their lives. Hence if welfare is poor because of inadequate housing, this is worse for the animal than some painful but brief interlude. A measure of how poor welfare is, multiplied by the duration of that poor welfare gives an indication of the overall magnitude of the problem for that individual (Broom & Johnson, 1993). Hence, the worst-case scenario would be severe prolonged problems.

If the welfare of individuals is poor, the more animals which are affected, the greater the problem. The largest numbers of animals with which man interacts are of farm animals, but there are also substantial numbers of pest species, pet animals, hunted animals, laboratory animals and zoo animals.

In the light of the considerations detailed above, and taking into account the fact that there are more chickens than any other domestic animal, the most extreme welfare problem areas would seem to be leg problems in broiler chickens and the consequences of confinement in laying hens kept in battery cages. Next, would come confinement of sows in stalls and tethers and of calves in small crates, disease in young calves and piglets, disease in farmed trout, mastitis and lameness in dairy cows and leg problems in turkeys and sheep. The inadequacy of stimulation possibilities in housing conditions for laboratory rodents and rabbits is an important problem, whilst injurious laboratory procedures on animals would be important although at a lower level than the problems resulting from housing. Before coming to laboratory procedures, we should consider the effects of transport on farm animals, of untreated disease in pet animals and of painful trapping or shooting of wild animals including fish. Farm operations such as castration, tail-docking or dehorning usually have severe effects as do tail-docking or other widespread mutilations in dogs. Neglect of pet animals and deliberate cruelty to domestic animals come quite low down on this list of problems. Even lower come housing conditions for zoo or circus animals and the consequences of damaging training methods for horses, dogs and other species.

This list of the severity of problem areas in animal welfare is based on the animals' point of view and is not the same order as that which most people would perceive. It is clearly desirable that people should be informed about the importance of the animal welfare, how to assess it scientifically and about where the really severe problems exist.

References

Arey, D. S. (1992). Straw and food as reinforcers for prepartal sows. *Applied Animal Behaviour Science*, **33**, 217–26.

Benus, I. (1988). Aggression and coping. Differences in behavioural strategies between aggressive and non-aggressive male mice. PhD thesis, University of Groningen.

Brantas, G. C. (1980). The pre-laying behaviour of laying hens in cages with and without laying nests. In *The Laying Hen and its Environment*, ed. R. Moss, *Current Topics in Veterinary Medicine and Animal Science*, vol. 42, pp. 129–32.

Broom, D. M. (1981). *Biology of Behaviour*, Cambridge: Cambridge University Press, 325pp.

Broom, D. M. (1983). The stress concept and ways of assessing the effects of stress in farm animals. *Applied Animal Ethology*, **1**, 79.

Broom, D. M. (1986). Indicators of poor welfare. *British Veterinary Journal*, **142**, 524–6.

Broom, D. M. (1987). Applications of neurobiological studies to farm animal welfare. In *Biology of Stress in Farm Animals: An Integrated Approach*, ed. P. R. Wiepkema and P. W. M. van Adrichem, *Current Topics in Veterinary*

Medicine and Animal Science, vol. 42, pp. 101–10. Dordrecht: Martinus Nijhoff.

Broom, D. M. (1988). The scientific assessment of animal welfare. *Applied Animal Behaviour Science*, **20**, 5–19.

Broom, D. M. (1989). Ethical dilemmas in animal usage. In *The Status of Animals*, ed. D. Paterson & M. Palmer, pp. 80–6. Wallingford: CAB International.

Broom, D. M. (1990). The importance of measures of poor welfare. *Behavioural and Brain Sciences*, **13**, 14.

Broom, D. M. (1991*a*). Animal welfare: concepts and measurement. *Journal of Animal Science*, **69**, 4167–75.

Broom, D. M. (1991*b*). Assessing welfare and suffering. *Behavioural Processes*, **25**, 117–23.

Broom, D. M. (1993). A usable definition of animal welfare. *Journal of Agricultural and Environmental Ethics*, **6**, 15–25.

Broom, D. M. (1998). Stress, welfare and the evolution of feelings. *Advances in the Study of Behavior*, **27**, in press.

Broom, D. M. & Johnson, K. G. (1993). *Stress and Animal Welfare*. London: Chapman and Hall.

Cronin, G. M. & Wiepkema, P. R. (1984). An analysis of stereotyped behaviours in tethered sows. *Annales de Recherches Vétérinaires*, **15**, 263–70.

Duncan, I. J. H. , Beatty, E. R. , Hocking, P. M. & Duff, S. R. I. (1991). Assessment of pain associated with degenerative hip disorders in adult male turkeys. *Research in Veterinary Science*, **50**, 200–3.

Duncan, I. J. H. & Petherick, J. C. (1991). The implications of cognitive processes for animal welfare. *Journal of Animal Science*, **69**, 5017–22.

Fraser, A. F. & Broom, D. M. (1990). *Farm Animal Behaviour and Welfare*. Wallingford: C.A.B. International.

Fraser, D. (1993). Assessing animal well-being: common sense, uncommon science. In *Food Animal Well-being*, pp. 37–54. West Lafayette, Indiana: USDA and Purdue University Press.

Gentle, M. J. (1986). Neuroma formation following partial beak amputation (beak-trimming) in the chicken. *Research in Veterinary Science*, **41**, 383–5.

von Holst, D. (1986). Vegetative and somatic components of tree shrews' behaviour. *Journal of the Autonomic Nervous System, Suppl.* 657–70.

Hughes, B. O. & Black, A. J. (1973). The preference of domestic hens for different types of battery cage floor. *British Poultry Sciences*, **14**, 615–19.

Hutson, G. D. (1989). Operant tests of access to earth as a reinforcement for weaner piglets. *Animal Production*, **48**, 561–9.

Knowles, T. G. & Broom, D. M. (1990). Limb bone strength and movement in laying hens from different housing systems. *Veterinary Record*, **126**, 354–6.

Koolhaas, J. M. , Schuurmann, T. & Fokema, D. S. (1983). Social behaviour of rats as a model for the psychophysiology of hypertension. In *Biobehavioural Bases of Coronary Heart Disease*, ed. T. M. Dembrowski, T. H. Schmidt & G. Blumchen, pp. 391–400. Basel: Karger.

Mansell, C. E., Elliott, H., Morris, T. H. & Broom, D. M. (1996). The use of a novel operant test to determine, the strength of preference for flooring in laboratory rats. *Laboratory Animal*, **30**, 1–6.

Marchant, J. N. & Broom, D. M. (1994). Effects of housing system on movement and leg strength in sows. *Applied Animal Behaviour Science*, **41**, 275–6.

Marchant, J. N. & Broom, D. M. (1996). Effect of dry sow housing conditions on muscle weight and bone strength. *Animal Science*, **62**, 105–13.

Mendl, M. , Zanella, A. J. & Broom, D. M. (1992). Physiological and reproductive correlates of behavioural strategies in female domestic pigs. *Animal Behaviour*, **44**, 1107–21.

Moberg, G. P. (1985). Biological response to stress: key to assessment of animal well-being? In *Animal Stress*, ed. G. P. Moberg, pp. 27–49. Bethesda, MD: American Physiological Society.

Norgaard-Nielsen, G. (1990). Bone strength of laying hens kept in alternative system, compared with hens in cages and on deep litter. *British Poultry Science*, **31**, 81–9.

van Rooijen, J. (1980). Wahlversuche, eine ethologische Methode zum Sammeln von Messwerten, und Haltungseinflusse zu erfassen und zu beurteilen. *Aktuelle Arbeiten zur artgemässen Tierhaltung, K. T. B. L. – Schrift*, **264**, 165–85.

Tannenbaum, J. (1991). Ethics and animal welfare: the inextricable connection. *Journal of the American Veterinary Medical Association*, **198**, 1360–76.

Vestergaard, K. (1980). The regulation of dustbathing and other behaviour patterns in the laying hen: a Lorenzian approach. In *The Laying Hen and its Environment*, ed. R. Moss, *Current Topics in Veterinary Medicine and Animal Science*, vol. 8, pp. 101–13. The Hague: Martinus Nijhoff.

10

New perspectives on the design and management of captive animal environments

DAVID SHEPHERDSON

The past decade or two has seen a revolution in the way in which we perceive the needs of animals in captivity. Nowhere are these changes seen more clearly than in the husbandry and design of enclosures for zoo animals. Exhibits in the past were seen merely as showcases for the exhibition of animal collections. Currently, they are increasingly envisaged as miniature habitats resembling the wild habitat of their occupants in which animals are encouraged to exhibit natural behaviours. The husbandry and design techniques used to achieve this are often referred to as environmental enrichment.

Although the primary motivation for environmental enrichment has stemmed from a desire to improve the welfare of zoo animals, there are many other benefits to be had. Many studies have shown that active animals performing natural behaviours within naturalistic settings are more attractive to the public and are perceived more favourably by zoo visitors (Sommer, 1972; Rhoads & Goldsworthy, 1979; Tripp, 1985; Shettel-Neuber, 1988), thus increasing the impact of conservation education. Increased reproductive success of endangered species has also been attributed to environmental enrichment (Dahl, 1982; Frisch, 1987) and it is likely that reintroduced animals emanating from more enriched captive environments will have higher rates of survival in the wild (Beck et al., 1986; Dietz et al., 1987; Shepherdson, 1994).

Life in the wild is full of hard choices. Feeding, grooming and travelling between resources such as nesting sites, food and mates are just a few of the wide range of behaviours that wild animals must employ each day of their lives in order to survive and reproduce. For most, time is limited and compromises must be struck. Too much time spent feeding may not allow enough time socializing with other group members or exploring for future food sites. The penalty for miscalculation can be severe and many pay the ultimate price, often after many days of suffering from lack of food, injury or disease. By comparison, life in a zoo has certain advantages. Food is in plentiful supply, travel distances are small, health is carefully monitored and there are no predators. In contrast to their wild counterparts, however, captive animals are faced with the problem of how to fill large amounts of time with the limited number of appropriate behaviours allowed by the captive environment. They also have little choice or control over their interactions with their captive environment.

Evidence that captivity can have an adverse effect on the behaviour of zoo animals is well established. Hediger (1955), Meyer-Holzapfel (1968) and Morris (1964) all described a number of abnormal behaviours such as stereotypic pacing, self-mutilation and apathy that appeared to be induced by captive environments and many subsequent studies have confirmed their observations (Fentress, 1976; Akers & Schildkraut, 1985; Chamove 1989). These observations are complemented by the results of so-called 'deprivation' experiments on laboratory animals in which the effects of removing specific environmental stimuli from animals under controlled conditions have been assessed. Harlow and Harlow (1962) documented a wide range of abnormal behaviours from rocking to self-mutilation that arose as consequence of depriving animals of social stimulation with conspecifics during their developmental period. Odberg (1987) showed that bank voles (*Clethrionomys glareolus*) were more likely to develop stereotypies when deprived of environmental complexity. Studies on the development of abnormal behaviours common in farm animals kept in intensive production systems also serve to illustrate the behavioural effects of stimulus deprivation (Fergusson, 1968; Kiley-Worthington, 1977; Dantzer & Mormede, 1983; Wiepkema *et al.*, 1983; Broom, 1988).

The degree to which abnormal behaviour can be used as an indicator of animal welfare, defined by Broom (1986) as the state of an individual with respect to its attempts to cope with its environment, is a hotly debated issue. While few would quibble with the notion that an animal displaying self-injurious behaviour such as tail-biting was suffering, the significance of some other behaviours, such as repetitive locomotor patterns, are not self-evident as Mason (1991) has clearly pointed out. What is needed is an objective measure which can tell us about the internal state of an animal. Unfortunately, no single measure exists. Instead, we have to rely on a multi-disciplinary approach in which welfare is inferred from an animal's health, physiology, behaviour and knowledge of its wild lifestyle. These sorts of studies are beginning to reveal a connection between captive behaviour and other welfare indicators. For example, Carlstead, Brown and Strawn (1993) found that, when caged domestic cats were subjected to a stressful management routine, urinary cortisol levels were elevated and the cats spent less time exploring and playing, and more time alert and attempting to hide.

However, there is another reason why the opportunity to perform natural behaviour patterns may be important to the welfare of captive animals and this is encapsulated in the concept of 'behavioural needs'. Animals may have a need to perform certain behaviours regardless of the usual purpose or goal of that behaviour. Mice, given the opportunity, will work to gather nesting material even when they have a fully functional nest (Roper, 1976) as will canaries (Hinde & Steel, 1972). Many animals will continue to work for food in the presence of free food (Neuringer, 1969). After being deprived of the opportunity to perform certain behaviours, such as wing flapping by hens (Nicol, 1987), many animals will show increased rates of that behaviour when they are finally allowed to perform them (referred to as the 'rebound' effect). The evidence for behavioural needs has been reviewed by Hughes and Duncan (1989). They concluded that motivation to perform behaviour does indeed exist. Thwarting a strong motivation may result in frustration which, in turn, can be a cause of suffering. However, two important problems remain, namely, how great is the motivation to perform different behaviours (and therefore what is the potential for frustration) and does the performance of abnormal behaviours satisfy the motivation. Further research may shed light on these questions.

Attempts to modify zoo environments in order to avoid the problems of captivity are not new. Robert Yerkes (1925) noted that '...the greatest possibility for improvement in our provision for captive primates lies in the invention and installation of apparatus which can be used for play or work'. However, in many cases the desire for easy animal management and strict hygiene requirements resulted in the traditional style of 'hard' zoo architecture consisting of concrete floors, iron bars and ceramic tiled walls. Whilst these enclosures allowed animals to be kept physiologically healthy, the psychological needs of the animals were often largely ignored. It would be a mistake to assume that this reflected a lack of concern for the welfare of captive animals. Rather, it was a combination of a different emphasis or appreciation of the relative importance of their needs and the quality of veterinary care and knowledge available at the time. An increased awareness of the complex perceptual and cognitive world in which animals live, coupled with advances in veterinary knowledge has, over the last decade or so, spurred a renewed interest in the improvement of zoo and other captive environments.

What then constitutes enrichment? One of the most conspicuous, and most remarked upon, differences between wild and captive environments is the amount of space available to the animals. Simply increasing cage size may be thought of as a form of enrichment. Small cage size has been linked to the development of stereotypic behaviour (Draper & Bernstein, 1963; Keiper, 1978; Keulen-Kromhout, 1978; Houpt, 1987), reduced reproduction (Glatston *et al.*, 1984; Snowdon, Savage & McConnell, 1984; Boot, Leusink & Vlug, 1985) and social instability (Sullivan & Paquet, 1977), so inadequate cage size clearly can present a welfare problem. However, as many authors have pointed out (e.g. Hediger, 1955; Shepherdson, Roper & Lups, 1990), the amount of space that wild animals use is determined by the dispersion of essential resources such as food, water and mates and not by a specific requirement for a given amount of space. A large number of studies have demonstrated that a similar principle applies in captivity. For example, Wilson (1982) showed that the presence of manipulable objects was a better predictor of activity than cage size for great apes. Berkson, Mason and Saxon (1963) found that cage size was not as important, with respect to stereotypic behaviour in chimpanzees, as the opportunity to engage in other behaviours such as object manipulation. Similarly, Odberg (1987) found that environmental complexity had a greater influence on the development of stereotypic behaviour in voles than cage size. Indeed, Erwin (1985) has described circumstances in which increases in cage size can be detrimental.

It would seem then, that beyond a certain minimum, the importance of space is indirect and operates through its effect on other variables such as temporal and spatial complexity, novelty, behavioural opportunities, availability of cover, proximity to other animals (from cage-mates to man), and the degree of control that an animal has in interacting with its environment.

In other words, it is quality rather than quantity of space that is important. However, the larger a cage is the easier it will be to achieve a given level of quality. It may be virtually impossible to provide realistic levels of complexity and change for a small primate in a laboratory cage but quite easy in a 10 Ha enclosure. Chamove (1989) has suggested that psychological space is a more useful concept than physical space. Chamove suggests that much environmental enrichment can be seen in terms of increasing the psychological space available to captive animals.

This is a similar idea to Hediger's (1955) space/time system. Psychological space can be increased in a number of ways, for example, by making more of the existing space usable for the occupant (e.g. by more fully utilizing the three-dimensional space with travel routes or by covering walls with climbable mesh), by requiring the animal to use more space to find hidden food, or by dividing the space with visual barriers.

There have been two main approaches to environmental enrichment, namely, the artificial approach pioneered by Markowitz at the Portland Zoo, Oregon, USA (Mellen, Stevens & Markowitz, 1981; Markowitz, 1982) and the naturalistic approach epitomized by Hancox (1980). Neither approach is exclusive to the other as Forthman-Quick (1984) has pointed out.

Artificial approaches can be subdivided into those that attempt to be functionally equivalent to natural tasks such as artificial 'termite' mounds and puzzle feeders and those that make no pretence at simulating a natural task but encourage behaviours or activities considered beneficial by other criteria (e.g. increased exercise). An example of the latter approach, often referred to as 'behavioural engineering', typically requires an animal to perform a task in response to an artificial stimulus, such as a light or noise. This approach has its roots in the experimental psychologist's technique of operant conditioning (e.g. Roper, 1976), whereupon receipt of a reward, usually food, is contingent upon correct performance of a task. For example, Markowitz (1982) developed a system for lar gibbons (*Hylobates lar*) which required that, whenever a stimulus light came on, the animals could operate two widely separated levers within a short time by brachiating between them in order to deliver a food item. The success of this device was measured in terms of increased activity/exercise and beneficial changes in social behaviour. In another example, a group of Diana monkeys (*Cercopithecus diana*) were trained to pull a lever to obtain plastic tokens which could then be inserted into a food dispenser in another part of the cage to obtain a reward. Again, increased activity, reduced aggression and interesting and varied social interactions resulted from this device. For example, some individuals would store up their tokens before claiming a food reward and would even use them to deliver food for their mother who had not learnt to use the apparatus.

The naturalistic approach aims to stimulate natural behaviour patterns spontaneously by providing natural stimuli or substrates. Thus foraging might be encouraged by scattering food onto a naturalistic substrate. The animals will then search for, and obtain, food in essentially the same way as they would in the wild. This method does not exclude the use of machines to deliver the food in the first place.

Both methods clearly have a place and are relevant in different situations. Behavioural engineering can be particularly effective in the highly artificial situations found in laboratory housing (e.g. Clarke *et al.*, 1987), whereas naturalistic methods may be more appropriate to the generally larger zoo enclosures. Whatever method is used, it is essential that enrichment should be both effective and practical: it must also be reliable and make efficient use of keepers' time and the financial resources of the institution. In a zoo this usually tends to favour the use of the simpler more naturalistic approach to enrichment.

That enrichment need neither be expensive nor time consuming has been demonstrated many times. Anderson and Chamove (1984) demonstrated how scattering small food items on to a straw substrate could enrich primate environments very effectively. Wilson (1982) conducted a survey which illustrated the beneficial effects of different types of objects in ape enclosures. More recently, Smith, Lindburg and Vehrencamp (1989) showed that, by simply providing macaques with whole fruit in place of chopped (a technique that also reduces preparation time), it was possible to increase the amount of time spent eating, the diversity, and the amount of food eaten. Carlstead, Seidensticker and Baldwin (1991) effectively enriched the behaviour of bears by hiding food in logs, boomer balls and soil. Similar results have been obtained using simple puzzle feeders (Shepherdson, 1988; Gilloux, Gurnell & Shepherdson, 1992) and insect dispensing devices (Shepherdson, Brownback & Jayes, 1989*a*). Even playback of conspecific vocalizations from the wild has been used successfully to enrich captive environments (Shepherdson, 1989).

Despite the ease with which these techniques can be initiated, it remains the case that more is written and talked about enrichment than is practised. We can only hope that those responsible for the care and maintenance of captive animals in future will pay more than lip service to the need to utilize these techniques.

References

Akers, J. S. & Schildkraut, D. S. (1985). Regurgitation/reingestion and coprophagy in captive gorillas. *Zoo Biology*, **4**, 99–109.

Anderson, J. R. & Chamove, A. S. (1984). Allowing captive primates to forage. In *Standards in Laboratory Animal Management*. pp. 253–6. Potters Bar, UK: Universities Federation for Animal Welfare.

Beck, B. B. , Kleiman, D. G. , Dietz, J. M. , Castro, M. J. & Rettberg-Beck, B. (1986). Preparation of golden lion tamarins for re-introduction to the wild. *Primate Report*, **14**, 198.

Berkson, G. A. , Mason, W. & Saxon, S. V. (1963). Situation and stimulus effects on stereotyped behaviours of chimpanzees. *Journal of Comparative Physchology*, **56**, 786–92.

Boot, R. , Leusink, A. B. & Vlug, R. F. (1985). Influence of housing conditions on pregnancy outcome in cynomologus monkeys (*Macaca fascicularis*). *Laboratory Animals*, **19**, 42–7.

Broom, D. M. (1986). Indicators of poor welfare. *British Veterinary Journal*, **142**, 524–6.

Broom, D. M. (1988). Sow welfare indicators. *Veterinary Record*, **123**, 235.

Carlstead, K. , Seidensticker, J. & Baldwin, R. (1991). Environmental enrichment for zoo bears. *Zoo Biology*, **10**, 3–16.

Carlstead, K. , Brown, J. L. & Strawn, W. (1993). Behavioural and physiological correlates of stress in laboratory cats. *Applied Animal Behaviour Science*, **38**, 143–58.

Chamove, A. S. (1989). Assessing the welfare of captive primates. In *Laboratory Animal Welfare Research: Primates.* pp. 39–51. Potters Bar, UK: Universities Federation for Animal Welfare.

Clarke, A. S. , Line, S. W. , Ellman, G. & Markowitz, H. (1987). Hormonal and behavioural responses of rhesus macaques to an environmental enrichment apparatus. *American Journal of Primatology*, **12**, 335–6.

Dahl, J. F. (1982). The feasibility of improving the captive environments of the Pongidae. *American Journal of Primatology*, **1**, 77–85.

Dantzer, R. & Mormede, P. (1983). Stress in farm animals: a need for re-evaluation. *Journal of Animal Science*, **57**, 6–18.

Dietz, J. M. , Castro, M. I. , Beck, B. B. & Kleiman, D. G. (1987). The effects of training on the behaviour of captive-born golden lion tamarins reintroduced into the natural habitat. *International Journal of Primatology*, **8**, 425.

Draper, W. A. & Bernstein, I. S. (1963). Stereotyped behaviour and cage size. *Perceptual and Motor Skills*, **16**, 231–4.

Erwin, J. (1985). Environments for captive propagation of primates. In *Primates: The Road to Self-sustaining Populations*, ed. K. Benirske, pp. 298–305. Springer-Verlag.

Fentress, J. C. (1976). Dynamic boundaries of patterned behaviour: interaction and self organisation. In *Growing Points in Ethology*, ed. P. P. G. Bateson & R. A. Hinde, Cambridge: Cambridge University Press.

Fergusson, W. (1968). Abnormal behaviour in domestic birds. In *Abnormal Behaviour in Animals*, ed. M. W. Fox, pp. 188–208. Philadelphia: Saunders.

Forthman-Quick, D. L. (1984). An integrative approach to environmental engineering in zoos. *Zoo Biology*, **3**, 65–78.

Frisch, D. J. (1987). Innovative designs for a lowland gorilla habitat. *American Journal of Primatology*, **12**, 343.

Gilloux, I. , Gurnell, J. & Shepherdson, D. J. (1992). An enrichment device for great apes. *Animal Welfare*, **1**, 279–89.

Glatston, A. R. , Geilvoet-Soetman, E. , Hora-Pecek, E. & Van Hooff, J. A. R. A. M. (1984). The influence of the zoo environment on the social behaviour of groups of cotton-topped tamarins. *Saguinus oedipus oedipus. Zoo Biology*, **3**, 241–53.

Hancox, D. (1980). Naturalistic solutions to zoo design problems. *Zoo Design 3, Proceedings of the 3rd International Symposium of Zoo Design and Construction.* pp. 166–73. Paignton Zoological and Botanical Gardens, Paignton, Devon, UK.

Harlow, H. F. & Harlow, M. K. (1962). Social deprivation in monkeys. *Scientific American*, **207**, 137–46.

Hediger, H. (1955). *Psychology of Animals in Zoos and Circuses*. London: Butterworths Scientific Publications.

Hinde, R. A. & Steel, E. (1972). Reinforcing events in the integration of canary nest-building. *Animal Behaviour*, **20**, 514–25.

Houpt, K. A. (1987). Abnormal behaviour. In *Veterinary Clinics of North America: food practice*. vol 3 (2) pp. 357–67.

Hughes, B. O. & Duncan, I. J. H. (1989). The notion of ethological 'need', models of motivation and animal welfare. *Animal Behaviour*, **36**, 1696–707.

Keiper, R. R. (1978). Environmental and physiological determinants of stereotyped behaviour in birds. *Proceedings of the 1st World Congress on Ethology and Applied Zootechnics (Madrid)*. vol. 1/2, pp. 495–7.

Keulen-Kromhout, G. V. (1978). Zoo enclosures for bears: their influence on captive behaviour and reproduction. *International Zoo Yearbook*, **18**, 177.

Kiley-Worthington, M. (1977). *Behavioural Problems of Farm Animals*. Stocksfield, UK: Oriel Press.

Markowitz, H. (1982). *Behavioral Enrichment in the Zoo*. New York: Van Nostrand Reinhold Co.

Mason, G. J. (1991). Stereotypies and suffering. *Behavioural Processes*, **25**, 103–15.

Mellen, J. D. , Stevens, V. J. & Markowitz, H. (1981). Environmental enrichment for servals, Indian elephants and Canadian otters at Washington Park Zoo, Portland. *International Zoo Yearbook*, **21**, 196–201.

Meyer-Holzapfel, M. (1968). Abnormal behaviour in zoo animals. In *Abnormal Behaviour in Animals*, ed. M. W. Fox, pp. 476–504. Philadelphia: Saunders.

Morris, D. (1964). The response of animals to a restricted environment. *Symposia of the Zoological Society of London*, **13**, 99–118.

Neuringer, A. J. (1969). Animals respond to food in the presence of free food. *Science*, **166**, 399–401.

Nicol, C. J. (1987). Behavioural responses of laying hens following a period of spatial restriction. *Animal Behaviour*, **35**, 1709–19.

Odberg, F. (1987). The influence of cage size and environmental enrichment on the development of stereotypies in bank voles (*Clethrionomys glareolus*). *Behavioural Processes*, **14**, 155–73.

Rhoads, D. L. & Goldsworthy, R. J. (1979). The effects of zoo environments on public attitudes toward endangered wildlife. *International Journal of Environmental Studies*, **13**, 283–7.

Roper, T. J. (1976). Self-sustaining activities and reinforcement in the nest building behaviour of mice. *Behaviour*, **54**, 1-2:40–58.

Shepherdson, D. J. (1988). The application of behavioural enrichment in zoos. *Primate Report*, **22**, 35–42.

Shepherdson, D. J. (1994). The role of environmental enrichment in the breeding and reintroduction of endangered species. In *Creative Conservation: Interactive Management of Wild and Captive Animals*, ed. G. Mace, P. Olney, & A. Feistner, pp. 167–77. London: Chapman and Hall.

Shepherdson, D. J. , Brownback, T. & James A. (1989*a*). A mealworm dispenser for the Slender-tailed Meerkat (*Suricata suricata*) at London Zoo. *International Zoo Yearbook*, **28**, 268–71.

Shepherdson, D. J. , Bemment, N. , Carman, M. & Reynolds, S. (1989*b*). Auditory enrichment for Lar gibbons (*Hylobates lar*) at London Zoo. *International Zoo Yearbook*, **28**, 256–60.

Shepherdson, D. J. , Roper, T. J. & Lups, P. (1990). Diet, food availability and foraging behaviour of badgers (*Meles meles L.*) in Southern England. *Zeitschrift Saugetierkunde*, **55**, 81–93.

Shettel-Neuber, J. (1988). Second and third generation zoo exhibits. A comparison of visitor, staff, and animal responses. *Environment and Behaviour*, **20**, 452–73.

Smith, A. , Lindburg, D. G. & Vehrencamp, S. (1989). Effect of food preparation on feeding behaviour of Lion Tailed Macaques. *Zoo Biology*, **8**, 57–65.

Smith, E. N. & Worth, D. J. (1979). Atropine effect on fear bradycardia of the eastern cottontail rabbit, *Sylvagus floridanus*. In *A Handbook in Biotelemetry and Radio Tracking*, ed. C. J. Amlaner & D. W. Macdonald, pp. 549–55. Oxford: Pergamon Press.

Snowdon, C. T. , Savage, A. & McConnell, P. B. (1984). A breeding colony of cotton-top tamarins. (*Saguinus oedipus*). *Laboratory Animal Science*, **35**, 477–80.

Sommer, R. (1972). What do we learn at the zoo. *Natural History*, **81**(7), 26–7/84–5.

Spurway, H. (1963). Can animals be kept in captivity? *New Biology*, **13**, 11–30.

Sullivan, J. O. & Paquet, P. C. (1977). Social systems of wolves in large and small enclosures. *Proceedings of the AAZPA Annual Conference*, pp. 207–16.

Tripp, J. K. (1985). Increasing activity in orangutans: provision of manipulable and edible materials. *Zoo Biology*, **4**, 225–34.

Wiepkema, P. R. , Broom, D. M. , Duncan, I. J. H. & Van Putten, G. (1983). Abnormal behaviours in farm animals. *Report of the Commission of the European Communities, Brussels.*

Wilson, S. F. (1982). Environmental influences on the activity of captive apes. *Zoo Biology*, **1**, 201–9.

Yerkes, R. M. (1925). *Almost Human*. New York: Century.

11

Should we let them go?

MARY MIDGLEY

Introduction

Is it wrong to keep animals in captivity? This is one of the questions often raised today, about the limits of our proper concern for animals. Many people now admit that we ought to have some such concern, but how far must it go? There is a natural wish for a definite terminus somewhere. The current search was presented in a television discussion not long ago. The philosopher Bernard Williams, when asked whether we owed any duty to animals, conceded at once that we do and said that it would be quite wrong to inflict pain on them. That, however, was where our responsibility ended. Nothing else (he said) could be expected of us. It would have been interesting to explore the reasons for this stance, to hear why extreme distress of other kinds mattered less than even mild pain. But, as does happen on these occasions, the talk veered off somewhere else. It may be worthwhile reflecting a bit further on the matter here. There seem to be three main reasons why many other people besides Williams might like to take this stance:

- People want a clear, definite terminus, and pain seems especially definite because it is physical. It strikes people as not having anything vague or subjective about it.
- Negative duties such as not inflicting pain seem limited and manageable, whereas positive ones such as producing happiness look infinite.
- This stance still allows the killing of animals – provided that it is painless.

How strong are these arguments? We might start by looking at the idea that physical pain is something perfectly definite and unmistakable, while mental distress is vague, undetectable and, in principle, quite separate from it. But, once we begin to think about this picture, it starts to look less clear. Neither animals nor people are really divided into two distinct

152

components – mind and body – independent of each other in a way that lets each suffer only its own special ills. The reason why people are not like this is that they too are animals – primates – and their central nervous systems are designed to make them function as a whole, not as two loosely connected components.

When a human being is harshly imprisoned, or is becoming quite exhausted, or is being sharply attacked, the mental and physical aspects of these misfortunes cannot be separated. All troubles beyond the mildest cause both kinds of damage. To make any animal (including us) thoroughly miserable is likely also to make it ill, as well as vice versa.

When we turn to the notions of objectivity and subjectivity, which scientifically minded people might hope to use to rescue us here, things grow even more confusing. In the first place, pain itself obviously has a subjective side. Notoriously, different people feel different stimuli differently and nobody else can really check on the difference. If we think of pain as objective, we mean that various outward signs, both in behaviour and in physiological functions such as blood pressure, reliably indicate it to outsiders – usually, in severe cases, and where there is no skilled pretence to mislead them.

Unfortunately, however, this same thing is also true of distress. The fact that you are in fear or in sorrow is objective even though, for you, this is a subjective event and its subjective quality is all important. The rest of us may be uncertain about your distress in certain cases, but we do not doubt that there is a fact there for us to be uncertain about. If we did, the idea of pretence could never have arisen. And, in most cases, the outside circumstances, along with the expressive signs, make it possible to know a good deal about the distress.

This however, is also true in the case of animals. Observers who are reasonably shrewd and experienced – which they need to be in human cases too – do not find it hard to identify most forms of grave distress in species of non-human creatures with which they are familiar. Unhappy animals pine, stop eating, creep away and become inactive, refuse to breed, attack those who interfere with them, become deranged, mutilate themselves, show stereotypic behaviours and die. Every efficient animal keeper learns to recognize these signs and to take them seriously.

Are we, however, ignorant of the causes of this distress? Is the psyche of other creatures perhaps an impenetrable mystery to us, as people often seem to say? On certain points it may be so. We certainly should not over-estimate our insight into it. Any species may have its own peculiar miseries and insensibilities, but the commonest sources of distress are not obscure to us in this way because we ourselves, as animals, are subject to most of them. It is agreed that, among these common causes of distress, close confinement can play a major role.

It is not actually easy to see how this could be doubted. When we watch a wolf or other active carnivore pacing to and fro in a narrow cage, or a bird of prey without space to fly, we are not presented with some bizarre, unintelligible enigma, as we might be by the behaviour of alien beings. In the first place, our own nervous system resonates at once with the actions of the frustrated animal, telling us what is wrong. And, when we cautiously doubt whether this is some anthropomorphic error, our full background knowledge of the species comes in to show us that it is not. We know that these are creatures which need much more space than they have for the movements proper to them.

Now, among the issues arising about our current treatment of animals, this particular trouble is outstanding. Space is expensive, so economics often seems to demand the closest possible crowding. Even in zoos this can happen, but its extremes are usually seen in food production – in pig and chicken housing, calf crates and lorry transport. Besides denying creatures their proper movement, this crowding radically insults natural social tendencies, which demand a certain minimum space between individuals. A similar, though converse insult attends solitary confinement, which is more common in laboratory conditions although it is often also found in zoos. Some experimentation involves close solitary confinement as part of its actual design, for instance, the stainless steel 'well of despair' in which baby monkeys were placed at birth for one series of Suomi–Harlow deprivation experiments (Suomi & Harlow, 1972), and the restraining harness used for dogs in experiments on 'learned helplessness'.

Is scepticism appropriate here? Ought we really to be in serious doubt about whether the animals are distressed? Is there some disreputable subjective element in the judgement that they are so? It is not easy to see what this element might be. It is true that we are judging something subjective. But then, we often have to do that, both for our own species and for others. For these occasions, the kind of objective evidence which we need is well known and consists of the expressive behaviour of the creatures concerned, along with our background knowledge of their species. In the case of the two exceptionally severe experiments just mentioned, it is of some interest that the expected outcome, and the whole point of the procedure, was to put the creature into a state of depression. Treatment designed for this purpose would not be expected to be pleasant to experience.

Confinement itself is not, of course, the only disagreeable factor in any of these cases, but we do not have to isolate it to show that it is such a factor. It is part of the case I am arguing that highly developed animals are such complex beings – so subtly integrated out of many interlocking systems – that it is scarcely possible for any serious trouble to strike them without also triggering off a host of others. The instrument resonates as a whole. This is notoriously one of the main difficulties about attempting simple control experiments in the psychology of motive, either for humans or animals, and it is also a central reason for being dissatisfied with the idea of attending only to physical pain when considering distress.

Suffering is then, a genuine, detectable, objective fact in the world – not a fantasy or an illusion – and confinement can be a genuine cause of it. How far should these considerations take us? Almost all keeping of animals – in farms, as pets or in zoos – involves some degree of confinement, and many traditional forms of it have been quite severe. (For the problem now confronting zoos, see McKenna *et al.*, 1987.) As we have begun to realize what this close confinement may mean, some people have started to object to all these customs. And, if we accept responsibility towards animals at all, as many of us now do, the issue does need to be looked at.

As usual, however, two extreme positions have emerged: one which defends all existing customs; another, equally sweeping, which wants to abolish them all. It will be worthwhile to look at both of them.

The defensive position

This often proceeds by dismissing all objection to confining animals as misconceived, sentimental and 'anthropomorphic'. Animals, it is claimed, feel no distress because they do not miss what they have never had. 'The modern hen', or calf is held to be quite changed from its more demanding ancestors, either because it has been spared the early conditioning or because it has been genetically changed so as to lose its inconvenient active tendencies.

Hens, however, when released from batteries turn out (after a period of shock and readjustment) to show exactly the behaviour of their unregenerate ancestors – scratching the ground, flapping their wings, brooding and socializing according to entirely acceptable chicken patterns. This makes great difficulties for the defensive case, whether that case is conceived as resting *a priori* on the general behaviourist principle that all activities must result from conditioning, or as an empirical claim that genetic interference has changed innate tendencies.

Behaviourist ideology is, however, still extremely powerful throughout the social sciences, even though most particular behaviourist doctrines have now been officially disavowed. Anyone who supposed it to have finally evaporated might be interested to read about its recent manifestations in Vicki Hearne's delightful book: *Adam's Task; Calling Animals By Name* (Hearne, 1987). The author, herself a professional animal trainer, points out how bizarre these doctrines look from the point of view of someone experienced in direct dealing with animals. But she shows how they still do dominate many circles where theorizing about animals goes on and is taken seriously by academics.

As these experienced people know well, animals do not have only a few 'basic needs' such as hunger and thirst, but possess very complex constitutions – sets of interlinked behavioural tendencies. Varied experiences during their lives can, of course, make these tendencies develop in very different ways, but they cannot remove them. These tendencies vary from individual to individual and also from species to species. Each set needs to be discovered empirically, and the errors that are called 'anthropomorphic' usually result from lack of this preliminary study.

That study, which starts from a general background of patient, unofficial observation by all sorts of people over the centuries, is now supplemented and articulated by the careful, methodical mode of scrutiny called ethology. Using its methods to survey these varying innate patterns, we notice that there are some animals for which confinement might be no injury at all, for instance, limpets or sloths. But, there are many others which are highly active and enquiring, so that they are bound to find monotony frustrating. There is nothing anthropomorphic about the observation that these creatures, like ourselves, are capable of boredom. The reason why we are capable of it is that we, too, are animals of this kind. Most birds and mammals fall into this class. Pigs, in particular, because of their opportunistic way of life are lively and inquisitive animals, quite unsuited for a life of close confinement.

Does it follow, then, that such animals demand unlimited freedom and are damaged by any kind of restraint? Interestingly enough, this does not seem to be true. Most animals that wander do so within fairly narrow limits. They tend to have a home range within which they feel that they belong; outside it they are uneasy. Many have also much smaller territories within this range to which they retire when alarmed and from which they do not want to be too distant. The notion of total vagrancy, 'roaming at random' which used to be credited to wild animals is an unreal one. Even migratory birds and far-ranging seabirds usually have a quite definite pattern of travel and like to know where they are going at any given time.

Again, then, the objections against confining animals turn out to be surprisingly close to those against confining human beings. Our own horror of being imprisoned does not mean that we want to roam indefinitely. We want a certain range, varying greatly with circumstances, and we want to know that we would be free to change it if we chose to. Once this is granted, many of us show patterns of travel much closer to those of the sloth and the limpet than those of the Arctic Tern or the Wandering Albatross.

If all this is true, how should we use it in our relations with animals? It surely suggests that, since we are not going to release them all tomorrow and wouldn't know where to put them if we did, we should start by trying to make the conditions of their confinement less nasty then they often are at present. With freedom as with many other matters, half a loaf is better, not worse, than no bread. Unquestionably, people who keep animals can often mitigate any distress that confinement may cause by attending to their natural tendencies, and can sometimes quite remove it. This is surely the first direction in which we should look for improvement, because it is often not hard to provide it. Pigs need a bit more space, some straw and a sanitary passage. Ingenious farmers willing to supply these things often manage to do so without going bankrupt. Given hard work and good will, piecemeal reform of this kind has repeatedly proved possible. Confinement has often been made more bearable by a bit of variety, for instance, by giving the pigs something to play with. And, as Konrad Lorenz pointed out, a caged bird that is let out of its cage every day for a good fly round is not noticeably worse off than many human commuters (Lorenz, 1961).

If one were an animal, already caught up in the system, one would surely be likely to vote for changes of this kind rather than no change at all. Some people, however, share Marx's contempt for half-loaves of this kind and regard them as dangerous 'reformism', a mere distraction from the deep revolution that is really called for. This protest brings us to the second main position we have to consider, namely that which calls for total abolition of all animal keeping.

Before leaving the defensive position, however, it may be worth remarking that it has not proved very impressive. It seems to require a kind of artificial scepticism about our knowledge of the inner life of animals – a scepticism which is unreal because it cannot be kept up if one is actually dealing with them. As Vicki Hearne points out, there are no animal trainers who do not attend with complete seriousness to the feelings of the animals they work with, because any who failed to would quickly end up dead.

Unreal scepticism of this kind is also sometimes professed about the feelings of human beings, including their pains, but it is hollow because it is not compatible with the practical conditions of life. We all have to take constant notice of the fact that we can hurt each other, and we are not wholly ignorant about when we are doing so. If, however, we turn to animals we notice that vets have to take this fact just as seriously as doctors do. Anaesthetics are needed for veterinary operations no less than for medical ones. If Descartes had been right – if animals had literally been unconscious automata – this close parallel would not be possible. And, what is true of pain is no less true of distress. Quite generally, the subjectivity of others is a central, objective fact of our life. If we can make sense of it in the case of other people, we can surely do so for other species too. Williams' idea of limiting our concern about it to physical pain in their case is not really workable.

What, however, about the second extreme position that we noted?

The abolitionist position

This forbids all confinement of animals. That veto is often supported today on the ground that such confinement is speciesist, that is, it shows an unjustifiable discrimination in favour of our own species over others. This discrimination is assimilated to sexism and racism as just one more indefensible and irrational example of unequal treatment. We have (it is claimed) no more rights to shut up animals against their will than we would have to do this to human beings.

The difficulty about this kind of appeal to equality is that we need to define the constituency – the group of equals – in order to make it bite, in advance, but it is just this definition that determines the conclusions (Midgley, 1983). How many kinds of animal are to be included here? Rats? Locusts? Mosquitoes? Most animals, after all, are insects. If it was arbitrary to discriminate between our own species and others, is it not equally arbitrary to discriminate between large animals which attract us and small animals which do not?

It is because these objections are so easily available to supporters of the (often odious) status quo that I greatly distrust arguments of this simple anti-discriminatory kind. They can, however, be proofed against some of these objections in order to carry their case a little further. Anti-speciesist arguments usually concede that one kind of difference is morally relevant in justifying discrimination – namely, difference in richness and complexity of life. If a locust's life is much less interesting than a dog's, then it may be justifiable to kill quite a lot of locusts to save some dogs. The locusts have less to lose.

This principle (like many utilitarian sub-principles) works conveniently to allow a good many conflicts to be settled in the traditional manner. It still lets us reach for the insecticide. But, to apply the principle, we have to represent the various commodities that are weighed against each other – life, happiness, pain – convincingly as countable and measurable. As with many other such utilitarian calculations, it is most obscure how we are supposed to have made this measurement. Might, for instance, the size of a swarm of locusts, or of a population of rats or gulls, make its joint sensibility outweigh that of a dog, or even the small human settlement it is about to overwhelm? Can we say of these animals, as Jeremy Bentham laid down for utilitarian decisions, that 'each is to count for one and nobody for more than one' (Mill, 1936)? In general, what are we to do about the fact that small and relatively simple animals like these immeasurably outnumber large and complex ones like ourselves or our dogs, while, though we may decide to count some of them as insensitive, we have really no clear grounds for doing so?

It seems very obscure how we could even begin to make the calculations about comparative suffering and enjoyment that would be needed for a genuinely egalitarian, 'non-speciesist' approach to the distribution of pain and pleasure. This would be a problem for God. Since we are neither omniscient nor omnipotent, we have to operate within the limits of what we know and can try to do.

To deplore this is bad faith; indeed, it is humbug. For we do not need this superhuman power and knowledge in order to see what we ought to be doing next. As things are, we are already doing very many things to animals which are so glaringly and unmistakably wrong that simply stopping doing them is an urgent duty. Many of these things fall well within the narrow limits proposed by Bernard Williams; they involve causing physical pain. Many more cause obvious and unmistakable distress – not only through confinement but through such things as isolation, terror and the separation of parents from young. If we start by remedying these obvious and shouting evils, we can then move gradually through slightly less disgusting country towards the cases that are really doubtful. It is only then that we would realistically face the question 'is it always wrong to confine animals?'

If we did face it, I do not think that we should get much mileage out of a simple egalitarian principle that we should not discriminate between different species. For one thing, what is good for different species genuinely does vary. There may be a duty to educate human children for human life but there is certainly not (what was at one time proposed for Washoe) any duty to educate chimpanzees for it by teaching them to talk. What the great apes need is to be secluded and held apart from human life. Their species distinctiveness is not some imaginary difference invented by prejudice. It is real. Respect for that distinctiveness is an essential part of any attempt to treat them decently. What they need is not 'human rights' but protection in their habitat – or, when they are in captivity, treatment that respects their species-specific nature.

Is there, however, some other reason, distinct from the doctrine of species equality, telling us that it is always wrong for people to keep animals? I do not find it obvious what this principle would be. Though there are many species which cannot conveniently live with people, there are a number of others which can and some which are actually co-adapted to do so. Of the many animals that have been successfully domesticated, ranging from cormorants to elephants, some, notably horses, pigs, dogs and cats, seem either to have been well suited to this life in the first place or to have been selected in a way that has made them so. When properly kept they can have a good life, and it is certainly a longer one on the whole than they could get on their own.

Is this advantage outweighed by loss of dignity? To say so seems to show a rather odd view of dignity. Working dogs are often very dignified, so are many other dogs. The notion that dignity demands complete independence – an autonomy only possible in solitude – is surely a distorted one in reference to human life and quite out of place in reference to dogs, who are not at all subject to the kind of arrogance that makes all dependence, whether material or emotional, count as an outrage on one's pride.

Altogether, the proper demand about these creatures is surely not that they should be got out of human life but that they should be given their proper place within it – a place which needs to be understood by reference to their special natures and capacities. Insults like the docking of dogs' tails and the declawing of cats show a crass disregard of those natures, as does the whole disgraceful business of breeding distorted animals for show. But then all this is bad pet-keeping, not pet-keeping as such.

What about zoos? In zoos there is a real conflict of interest because, in order to get their animals, zoos depend on trappers in the wild who are putting increasing pressure on more or less endangered species, disturbing habitat and killing many creatures for every one that they finally deliver.

This, however, is a different point from the treatment of those who are already there and cannot be released. For them, again, the question is surely not whether they are kept but how. In the last few decades, certain enterprising zoos, such as Gerard Durrell's zoo in Jersey, have made remarkable progress in devising quite uncage-like surroundings for their creatures. In such zoos, adaptable species, in suitable numbers and the right environment, appear to have quite a good life. No doubt it is not perfect, but then what life is perfect?

Is there a danger for them of a kind of corruption, a depraved dependence by the animals on their keepers? Durrell himself somewhere tells a nice story of an occasion when transport failed him on a collecting trip and he was forced to release most of the animals that he had collected. Next morning they were all back, waiting for breakfast. Understandably, Durrell saw this as a straightforward vindication of zoos. And, up to a point, it is so. But, of course, these animals were having it both ways. They were not reflectively signing a contract to abandon their liberty. They were not envisaging the kind of inactivity that is involved over a long term in not being able to earn your living – a problem that is particularly severe for captured carnivores.

All this is a real difficulty for zoos. But granted that these zoos are already there, the main point is surely that their staff should recognize this difficulty – that they should make efforts, as indeed many of them are now doing, to provide the most suitable life possible for each species in the conditions that are actually available. Television documentaries have, in fact, educated public taste on this matter to a point where traditional crude cages are beginning to strike many people as offensive. Decent zoos are therefore increasingly seeing the need to move towards keeping fewer species and giving them better conditions. There is something to be said for encouraging this move rather than merely deploring the whole institution and trying to abolish it. And *mutatis mutandis* – because the whole scene is much darker – the same thing may be true of experimentation.

A word finally about the kind of moral reasoning that is in place here, I have suggested that sweeping generalizations are not appropriate. This is a serious professional view, not a slovenly refusal to philosophize. Some moral philosophers tend to suggest that what is needed from philosophy on moral questions is always startling, highly abstract theories, often paradoxical in appearance, often very hard to apply in particular cases, and producing results quite unlike our existing moral judgements. These are beds of Procrustes, on which subjects are laid to be stretched or decapitated if they fail to fit.

Although drastic change is sometimes needed, this kind of approach gives philosophy a bad name, and rightly so. The reason why we need to philosophize is that our existing notions fall into conflict and confusion. When this happens, we have to start doing it by thinking harder about these existing notions, about the meaning of our current moral judgements and principles – not just by seizing on one element in them and backing it against all comers.

Sometimes this kind of reflection does lead us to form views that are more sweeping and more one-sided. For instance, we may conclude that all slavery or all torture is wrong, not only some forms of it. But, on other issues, this kind of sweeping result does not emerge. For instance, we do not necessarily have to conclude that all marriage must be abolished when we make divorce possible, nor that there should be no such thing as the family when we take steps to free people from domestic tyranny, although reformers have proposed both these drastic conclusions. Similarly, if we expect a sweeping, simple answer to the question, 'Is it right or wrong to keep animals?' we are likely to be disappointed. But then we know that, with other and more familiar questions, such simple answers are often fraudulent. Sweeping prohibitions like 'never lie' or 'never kill' fail because there are occasions when the alternative is worse than doing these things. The prohibitions have a strong point because they mark the strong general objections to these acts, but they cannot save us the trouble of weighing up considerations in exceptional cases. The idea of equality is an invaluable tool for cases where what has gone wrong is indeed an objectionable kind of inequality. The word rightly has a good name because its main use has been in political situations where that is actually true. But there are other situations where it works extremely strangely. For instance, it has sometimes been argued that we do wrong to be partial to our own children and other relatives. This thought has led some theorists (such as Plato, J. B. Watson (1928) and Shulamith Firestone (1979)) to suggest that we should get rid of this partiality by not letting people know who their own children or parents are at all. Others have said that we may know this, but still ought not to favour them above anybody else (Godwin, 1971). This hardly seems realistic. The special affection felt here is a natural one, without which most children would, throughout history, never have been brought up at all, simply because the trouble involved is excessive. Adoptive parents do not dispense with this partiality; they either develop it or stop bothering. Partiality is indeed wrong if it gets out of hand, for instance, if we start shooting other people's children or failing to provide through our community what is necessary for them or even not being normally nice to them when the occasion arises. Natural partiality is not a licence for writing anybody off. But it is a perfectly proper licence for feeling special affection and acting on it to provide special care. Anyone who supposes that equal affection should be felt for everyone, or even for everyone of equal merit, seems to be trying to treat people as abstract intellects instead of flesh-and-blood human beings.

This case surely supplies a useful parallel for the case of other species. In both cases, there is some natural limitation on sympathy. In general, we are no more capable of spreading our concern equally and universally to all species than we are of spreading it equally to all children. The one case is a more advanced form of the other, since animals have children too.

But, this is no kind of excuse for turning our back on non-human animals. We do not need this immensely ambitious form of egalitarianism in order to see that in many ways we ought not to be treating them as we do. Ordinary moral considerations such as the Golden Rule ('would you like that done to you?') show us that we have a duty to do immensely more to help and protect than we are now doing. It seems to be best to let such plain moral points speak for themselves, rather than to dress them up in extreme dramatic forms which those not yet converted will be much more likely to reject entirely.

References

Firestone, S. (1979). *The Dialectic of Sex: The Case For Feminist Revolution.* p. 222. London: The Women's Press.

Godwin, W. (1971). *Enquiry Concerning Political Justice.* Book II, Chapter 2, p. 71. ed. K. Cadell Carter. Oxford: Clarendon Press.

Hearne, V. (1987). *Adam's Task: Calling Animals by Name.* London: Heinemann.

Lorenz, K. Z. (1961). *King Solomon's Ring.* p. 72. Translated by Marjorie Kerr-Wilson. London: Methuen University Paperback.

Midgley, M. (1983). *Animals and Why They Matter.* chap. 6. Athens, Georgia: University of Georgia Press.

McKenna, V. *et al.* (ed.) (1987). *Beyond the Bars: The Zoo Dilemma.* London: Thorsons.

Mill, J. S. (1936). *Utilitarianism,* Chap. 5, p. 58. London and New York: Dent and Dutton.

Suomi, S. and Harlow, H. S. (1972). Depressive behavior in young monkeys subjected to vertical chamber confinement, *Journal of Comparative and Physiological Psychology,* **80**, 11–18.

Watson, J. B. (1928). *Psychological Care of Infant and Child.* pp. 5–6. New York: W. W. Norton and Co.

.

Part IV

Research and education

12

Humane education: the role of animal-based learning

ANDREW J. PETTO and KARLA D. RUSSELL

Introduction

The anthropologist Claude Lévi-Strauss (1965) is well known for his statement that human societies revere totemic animals not because they are good eating, but because they are good thinking.[1] His work points out many ways in which humans understand the world and their place in it through their knowledge of the animals around them. Indeed, this way of knowing the world through the animals around us may be considered one of the human 'universals' (cf. Brown, 1991).

Furthermore, human societies rely on this animal-derived information in several characteristic ways – symbolic, formative, cautionary, and observational – and we recognize these in the many ways animals are represented in our own culture. Löfgren (1985) describes the post-bourgeois development of the idea of 'animality' that viewed the animal world as degenerate and immoral, and established a hierarchy of the animal kingdom based on the tendency to express certain 'good' characteristics. In a similar vein, we know the cautionary tales of Aesop and Beatrix Potter in which the prominent characters are mice, foxes, or turtles and hares (Kale, 1993). These views became the basis for formative lessons about right and wrong.

Gillespie and Mechling (1987) show the elevation of certain salient characteristics of animals so that these animals stand as symbols of the character of a people or as mascots. In the US they trace the symbolic rise of the bald eagle and the concomitant decline of the wild turkey as an

---

[1] This statement by Lévi-Strauss is most often translated as 'animals are good to think', because in the original text, he contrasts the role of totemic animals as sources of knowledge about the world against their role as good to eat. 'On comprend enfin que les espèces naturelles ne sont pas choisies parce que "bonnes à manger" mais parce que "bonnes à penser" (1965, 128). The translation we use is thanks to Natalie Maynor, Professor of English, Mississippi State University.

example of this process. Trickster stories in which a smaller, weaker, or slower protagonist defeats a more powerful opponent by use of his wits are also common across cultures (e.g. Frey, 1987).

Finally, some use of knowledge about animals is derived strictly from observing their behaviours in nature. We mark the return of spring in the US by the appearance of the first robin, or the end of the summer by the sighting of flocks of migratory birds beginning to fly in formation. Some of this information may come to us through nature magazines or broadcast programmes, from newer communications media (CD-ROM and interactive video), or from formal lessons and programmes in schools, camps, and enrichment activities. Some of it is passed down through the generations directly or indirectly.

In all these ways, we learn and come to 'know' about the world through the animals we encounter in legend, in popular culture, and in nature. Lévi-Strauss argues that much of what we 'know' about animals is in the cultural or sociological domain, not in the natural or biological (Lévi-Strauss, 1965:90). More recent work by Hatano *et al.* (1993) further illustrates the premise that it is through the acquisition and sharing of this cultural knowledge that we form our attitudes and expectations about the relationships among human and non-human animals, their respective roles in the world, their needs and wants and our appropriate responses to them. It is also in this process that our concept of 'humane' emerges.

Concept of the 'humane'

'Humane' is a cognitive concept for humans and subject to the same constraints as other cognitive concepts held by humans (Atran, 1990). It is universal in the sense that all cultures seem to have a concept that some actions and attitudes toward animals (and toward other humans) are desirable and others are unacceptable. However, often the set of acceptable and proscribed actions towards non-human animals differs greatly from one culture to another, and, even within a culture, attitudes towards treatment of animals can vary by class or socio-economic status (Driscoll, 1992; Löfgren, 1985).

Within western cultures, various studies have explored the relationship between abusive treatment of animals and abusive behaviours toward other humans (e.g. Ascione, 1993; Felthous & Kellert, 1986; Kellert & Felthous, 1985). Conversely, Ascione (1992) reports that humane educational exposure to animals increases empathy in human-to-human interactions among children, and Weatherill (1993) also reports carry-over of caring behaviours toward classroom pets into other aspects of the children's learning. Finally, Katcher, Beck and Levine (1989) describe how the use of animals in an imprisoned criminal population can be used to foster a more humane attitude toward other humans.

The challenge for anyone trying to describe a process through which one learns about animals and with animals as 'humane education', then, is to focus not only on the final rules for behaviours toward animals, but also to examine the pathways to those rules. There are two main goals of this examination. The first is to find opportunities in the learning process to understand better both the 'natural' and the 'cultural' animal (sensu Lévi-Strauss, 1965) and to discover what we can learn from all the different ways in which our culture and others know these animals. The second is to refine the pedagogical process so that we develop in our students a humane attitude that includes appreciation of the animal's natural life, role in the environment, and the costs (to animals and humans) of its capture and study.

The process of considering these issues for animals in education has three stages. The first stage focuses on pedagogical issues and is generally the domain of the teacher. The main issues in this stage relate to the objectives of the lesson, integration of the animal-based activities with other aspects of the curriculum, the design and presentation of the materials, actions and reactions of the learners, and an evaluation of the learning by each individual as well as of the lesson or activity as a whole. The second stage focuses on the impact on the animals themselves. The main issues in this stage relate to the acquisition, care and use, and disposition of the animals being used for education. Finally, the third stage focuses on the wider social impact. The main issues in this stage relate to the outcome(s) of the process on the educational climate in the schools, the community, and in society in general.

We do not believe that this approach will or must lead to an abolition of animals in the classroom nor that it should do so. Rather, we hold that humane education is embodied in the process of considering a variety of issues including the nature of the lesson to be taught, the opportunities for multiple approaches to that knowledge, the active consideration of the life (sensu Regan, 1993) of the animal subject as an important issue, the conservation of resources, and the outcome of the exercise for the teacher, the student, and the animal.

Our use of the term 'active consideration' throughout this chapter is meant to convey a sense that each choice to use animals in education is explored and investigated by the teachers and students as appropriate to the students' experience and abilities; that this exploration is not merely a perfunctory checklist of health, safety, and physical comfort issues, but an integral part of the educational experience; that the conclusions and choices to be made are not a foregone conclusion before the process begins; and that executing this exploration requires a set of learning activities that may take the student beyond the immediate lesson and classroom environment to do background research, to check sources of information, to document past educational uses and their outcomes, etc. If successful, the members of the learning community – teachers and students – have turned the hit-or-miss experiences of the classroom 'pet' into an integrated, multi-disciplinary exploration of the biology, psychology, economics, and anthropology of educational use of non-human animals.

The experience of this process is the essence of humane education. The absence of real experience with non-human animals in the context of a humane educational setting eliminates an important opportunity to develop the concept of 'humane' in our students. Unless all members of the learning community are actively engaged in learning how information about animals is obtained and used in the classroom, we cannot fully demonstrate to the community in practice how to foster an environment of respect for those animals. We illustrate the values of humane education by accepting the responsibility to think clearly and responsibly about the role(s) that animals may play in our planned educational activities and the impact of those activities on the lives of animals.

Issues in teaching and learning

There are many ways in which animals may appear in an educational setting, but we will be concerned with just two subject areas – biological and behavioural sciences. We restrict our consideration to these two areas precisely because these are the areas of study in which one expects that the animals must be present in some form for the learning to be successful. Most other learning about animals in school is precisely that – learning about animals. However, it is conceivable that, in a fully integrated curriculum, one would include the observation of living animals in a unit on, say, animal folklore. We will also concentrate on the issues of bringing animals into the classroom or laboratory setting, but at least some of our examples, and our own professional preference, is to take the laboratory and the classroom to the animals where and how they live.

We also will take a broad view of education as encompassing pre-kindergarten through post-doctoral training, as there are always opportunities and needs to re-evaluate the educational use of animals at all these levels. However, most of the examples that we will use and most of the discussion will centre on secondary and introductory level university students. We believe that the approach and concepts apply through a lifelong education, but our examples drawn from our own teaching and learning experiences draw us to this more restricted phase in our student's formal education.

The first step in a humane approach to animals in education is for the teacher[2] to identify the best pathway to meet the lesson's objectives.[3] The teacher must take into account the learners' stage of cognitive development, prior or collateral knowledge that the learners bring to the lesson, resources available to plan and execute the lesson, plans for evaluating the success of the lesson and the learners, the internal environment of the classroom, and the external environment imposed by systemic or other standard for mastery of life sciences content and concepts at this and subsequent stages of education (see Table 12.1).

Once these elements are identified and arranged into a teaching plan, then the teacher can evaluate how the use of animals may contribute to the learning activities. Examples of this approach are described in the draft standards for science education published for comment recently by the National Research Council (NRC) of the US (NRC, 1994).

If the lesson or any learning activity will include the use of live animals or animal products, the teacher first should be able to provide a compelling and significant pedagogical justification for such use. That means that the use of animals in the classroom provides an added component to the learning that is non-trivial and unique or unattainable in other ways and that there is substantive evidence to support this assertion. This evidence may come from educational research on learning styles (e.g. Gardner, 1993), from practical guides to the implementation of this research (e.g.

[2] Throughout this chapter, we will use the term 'teacher' to refer to the individual responsible for setting the learning objectives, planning the lesson or the curriculum, executing the lesson, and evaluating both the performance of the learners and the extent to which the lesson has achieved its objectives. As such, 'teacher' may refer to any number of individuals involved in the process of education beyond the one who conducts any specific class or lesson. We recognize the teacher as the educational leader of the class and as a technical 'expert' in pedagogical issues, so we grant the teacher a leading role in the determination of pedagogical approach and justification. This leadership is, perhaps, more pronounced with younger children and changes in character as the learners develop more complex cognitive capacities. Although parents may serve as the audience for pedagogical justifications and rationales when children are young, it is not clear that they generally serve in this capacity in most schools in North America.

[3] This discussion may refer to any aspect of the curriculum – classroom lessons, learning activities, curriculum units, or even school- or system-wide guidelines.

Table 12.1. *Issues in teaching and learning with animals*

Learning styles (intelligence)
Does the proposed activity allow or encourage acquisition and construction of
    knowledge by learners in a variety of ways?
Does this proposed activity engage the learner actively in the process of
    discovery, learning, evaluation, and assimilation of knowledge?

(Cognitive) developmental stage/age/level
Is the proposed activity appropriate to the abilities of the learners to understand
    and assimilate the main points of the lesson?
Is the proposed activity better performed at an earlier or later developmental
    stage?
Has the prior preparation for the proposed activity been adequate and
    appropriate to both the developmental stage of the learners and to the expected
    learning outcomes or culminations?

Lesson objectives
When the lesson objectives and goals are clearly formulated, how does the
    proposed activity support their attainment? What skills or knowledge are being
    developed, and how and when are they necessary for future learning?
What other pathways to the objective might be used and how would they affect
    the educational outcome of the activities?
Will the proposed activity be a superficial, one-time event or will it reflect and
    support the main theme throughout a curriculum unit or longer-term
    educational effort?

Career stage
How does the development of specific skills and knowledge translate into a
    potential for future study or career choices for learners?
Conversely, how would lack of specific skills inhibit the student's future plans
    and expectations?
What are the best ways to learn these skills and to what depth at this stage in the
    learner's academic career?

Markova & Powell, 1992) or from research that specifically addresses
learning with animal-based lessons (e.g. Kinzie, Strauss & Foss, 1993). The
teacher may also rely on prior personal experience with animal-based
learning or the experience and advice of professional colleagues, but this
should have a firm pedagogical basis.

Most educational uses of live animals or animal tissues are based on the
demonstrated value of a practical or 'hands-on' component to the lesson.
The power of adding visual and 'bodily-kinaesthetic' components to what
Gardner (1993:8) called the 'linguistic' and 'logical-mathematical' biases
of 19th and 20th century education is illustrated in many disciplines. This
approach has been most appreciated, perhaps not surprisingly, in arts
education (e.g. Petto, 1994; Lowenfeld & Brittain, 1970; Arnheim, 1969).
In these disciplines, both learning and its evaluation take into account a

rich array of interactions among the teacher, the learner, and the subject matter, including sensory, emotional, spatial, interpersonal, and kinaesthetic.

There is no question that what educators call active learning throughout multiple modalities makes learning better in at least two ways. First, students learn more when they confront learning problems that engage them in inquiry, problem posing, problem solving, and defence of their ideas before their classmates (Peterson & Jungck, 1988; Jungck, 1985). In most cases, teachers are referring to their personal experiences as well as a reflection on their intuitive (emotional or interpersonal, sensu Gardner, 1993, 9) sense that hand-on laboratory activities with animals add significantly to learning biology (e.g. Offner, 1993; Keiser & Hamm, 1991; Mayer & Hinton, 1990). This is not merely a matter of developing manual dexterity or hand–eye co-ordination or facility and self-confidence with some laboratory technique, as some have described it (e.g. Kinzie *et al.*, 1993; Quentin-Baxter & Dewhurst, 1992). Rather, these practical or hand-on lessons provide non-linguistic ways of learning, and for some students the movement, proprioception, and emotional reaction to the learning and to other learners cannot be replaced by linguistic, visual or symbolic (i.e. logical-mathematical) representations of the problem.

Secondly, this approach to learning engages more students in the process (e.g. Petto, 1994; Gardner, 1993; Markova & Powell, 1992). Such a wider engagement allows more students to participate in, and contribute to, the learning experience and may give them more of a sense of control or self-direction in constructing their own learning. In addition, Petto (1994) reports that the personalization of the learning activity through the incorporation of the emotional response to the activity, materials, and even the other learners is a key factor in both the retention of learned material and the ability of students to relate that material or lesson to other knowledge or life experience. Furthermore, the perspectives of those students whose learning is not primarily linguistic or logical-mathematical contribute insights into the learning that may be overlooked by fellow learners, including the teacher. Even the learners who prefer expository teaching and declarative evaluation of their learning, learn more and better when using multiple modalities, as illustrated in a recent study on reinforcing the lessons learned through dissection by using prior preparation with an interactive video demonstration (Kinzie *et al.*, 1993).

Although the value of such hands-on learning in the biology laboratory is seldom challenged, critics charge that biology teachers are at least unimaginative in their approaches to animal use in education and, at worst, abusive. In the former vein, Texley (1992:25) argues that much of what drives animal usage, especially dissection, in the biology classroom is inertia and that '...the lack of training and preparation among many science teachers means that dissection is often the only hands-on experience that they know'. Furthermore, students may not be properly prepared for either the technical or the experimental aspects of learning through dissection. Mansbach and Simmonds (1986) argued, by analogy, that there are many other hand-on activities in the sciences that we could, but do not, let students carry out in school, such as constructing and detonating explosive compounds in chemistry class.

Finally, the main issue in humane education with animals is that biology is the study of the living (Lock, 1994; Orlans, 1991). In teaching and learning about living animals, one might consider, for example, their way of life, social and environmental needs (in nature and in captivity), feeding strategies and nutritional needs, and their role in the ecosystem. Such considerations are found in the proposed NRC draft standard for education in the life sciences (NRC, 1994), but are also common in the literature directed at the practising science teacher (e.g. Benham, 1991; LaHart, 1991; Bisbee, 1990; Beck, 1968). Some resources, like Bisbee (1990) or Field and Shapiro (1988), contain checklists that, in themselves, bring questions of the appropriate use of animals in education and the impact on the animals chosen into focus for teachers, administrators and students. This approach raises a paradox for the learning community, since almost any proposed educational use of animals will disrupt the animals' lives to varying extents.

One solution to this apparent paradox was proposed by Donnelley and colleagues (1990) under the term 'moral ecology'. Considering the moral ecology of the proposed use of an animal in education (or research) includes asking about the life that the animals (would) lead outside the educational context and how any proposed use would contribute to the educational objectives of the lesson. This is where the educational use of animals becomes humane. First, the teachers and students examine what needs to be learned and how an animal might contribute to that learning. Next, they review the needs of the animals and the impact on that animal of the proposed learning activity, considering, perhaps, alternatives that include using the animals in a different way, using different animals, or using non-animal resources. Then they should discuss the source of the animals, their acquisition, and their disposition after the educational activity.

These considerations are based on our knowledge of animals in the natural world, and the active consideration of these issues and acquisition of this knowledge play a powerful role in teaching both the biological and the ethical/humane lessons that we wish students to learn. This may occur simply because the teacher is creating the environment in which these questions are addressed (e.g. Lock, 1993; Mayer & Hinton, 1990; Mansbach & Simmonds, 1986).

Such a process places a heavy responsibility, however, on science teachers who may not have had any formal training in bioethics, particularly in exploring complex ethical issues with children (Downie, 1993; Downie & Alexander, 1989). Lock (1993, 114), in particular, points out the responsibility of the teacher to demonstrate 'a caring and humane approach in all their work with living things'. The responsibility for the teacher, then, is to be sure that the students have accurate, up-to-date information from a variety of sources about the animals they propose to study, including information relevant to the moral ecology of the use of particular animals in specific learning activities and projects.

In summary, the justification of any educational use of animals must have a strong pedagogical basis. This justification must include consideration of the choice of species, the type of learning activity, the developmental readiness and scholastic abilities of the students, the necessity for adequate foundations for future study, and advanced preparation and study by the students and teachers. A part of this justification is to balance the needs of, and outcomes for, the learners against the impact of the proposed educational usages on the animals.

Effects on animals

After careful consideration of the pedagogical issues, if the teacher concludes that there is an appropriate educational role for animals in the lesson, then s/he must determine whether there are any animals suitable for the lesson and the classroom environment. The main issues pertain to the acquisition, care and use, and disposition of the animals used in the lesson. Much of our discussion derives from the concept of 'moral ecology' developed by Donnelley and colleagues (1990) and later expanded in the context of animal biotechnology (Donnelley, 1994).

Moral ecology may be viewed as an attempt at operationalization of the 'subject-of-a-life' criterion proposed by Regan (1993). It presents a set of principles against which we might explore by what criteria we may judge the subjective lives of animals and the impact on their 'individual experiential welfare' (Regan, 1993, 203) of various uses of these animals by humans. Moral ecologists recognize that the life's experience and the expectations for future life differ greatly among individuals of the same species (e.g. Sapontzis, 1987; Rodd, 1990). Therefore, the impact upon their individual experiential welfare of their interactions with humans in an educational setting may also be different.

We can easily imagine how this could be the case. The experience and expectations for life of a purpose-bred pet hamster must be different from that of its wild conspecific cousin. The same holds true for the horse, for which both wild and domestic herds exist. Furthermore, horses kept for work, recreation, entertainment, or sports are all likely to have different experiences of their own lives. Therefore, the assessment of how educational use may affect them depends both on the proposed educational use and on the animal's current existence.

Perhaps the most important part of exploring the impact of the use of animals in education on the animals themselves is to expand our considerations of the issues of humane teaching and learning beyond the salient issues of classroom dissection or other learning activities that cause or require the death of animals. Although there is hardly any instance in which human observation of animals, whether casual or purposeful, does not have an impact on these animals (Lott, 1988; Davis & Balfour, 1992), there is reason to believe that some learning activities are tolerated more easily by these animals than others. However, a humane approach requires that everyone involved in the educational use of animals explores explicitly the effects of the proposed use on the animal subjects. Indeed, such background research before any classroom activity is a hallmark of the proposed standards for life sciences education from the US National Research Council (NRC, 1994).

Through the application of the framework of moral ecology, we can learn much about the behavioural ecology and the interrelationships among organisms in nature. One example from the draft standards for science education issued by the US National Research Council (NRC, 1994) explores the use of earthworms in a terrarium to illustrate the creation of environments within the school that more than adequately meet the environmental and social needs of invertebrates and even small vertebrates (e.g. Friedman, 1991; Wiessenger, 1990; Majerus, Kearns & Forge, 1989). Teachers can also take advantage of animals, such as spiders, flies, and ants that typically inhabit school buildings or those observable in the schoolyard (e.g. LaHart, 1991). Problems in keeping such animals in schoolrooms, such as those pointed out by Cantor (1992), are easily avoided by creating appropriate microhabitats within the larger schoolroom environment.

The issues in Table 12.2 expand the sphere of inquiry of the effects of educational use on the animals beyond whether the subject animals will live or die. This process includes learning about their lives before the animals come to the classroom (in nature or in any other environment), how the animals will be cared for and by whom, how the proposed use will affect the animals and the learners, and what the effects of this activity might be on the animal's future life once the project is over. It also requires us to identify and evaluate the sources of this information and to determine what message each of these is bringing to the lesson at hand.

Although no such list can ever include all the issues that could be raised, we believe that the process of considering explicitly the impact of educational usage on the animals themselves is vital to the development of a humane ethic in education. The desire to add other items to the list is a healthy expression of a learning community that takes seriously the need for such a development. Perhaps most importantly, this list is applicable to all animals and to any proposed use in education from behavioural observations of free-ranging animals in the schoolyard to dissection of mammalian species.

(Human) social issues

An important question in the use of animals in education that is often overlooked is the effect on human society and on the learners that experience it. Both the lore of scientific training and the criticisms from the animal rights literature point out the distancing, the deadening of emotion, the objectification of the animals, and the desensitization to suffering and death that educational uses of animals can have on the people who use them (e.g. Davis & Balfour, 1992; Shapiro, 1990, 1991). These emotional 'adaptations' are expected for all uses of animals, but are particularly pronounced when the animal use results in death or dismemberment or involves suffering. These studies argue that being forced to partake in these activities may require a psychological adjustment by the learners that degrades or devalues animal life. Similar reactions to the plight of human subjects in scientific research has been well documented for decades (e.g. Milgram, 1974). Under social pressure from peers and authority figures, experimental assistants new to the project were rather easily convinced to administer what they believed were painful procedures to unseen subjects for the sake of the experimental protocol.

However, the contributors to the volume by Davis and Balfour (1992) demonstrate that this outcome is not unavoidable. Furthermore, re-

Table 12.2. *Inventory of issues for use of animals in education*

Acquisition
    How are the acquisition and use of the animals to be introduced to the students?
    How and from where will the animals be acquired?
    Can they be studied in their natural habitats, or must they be introduced to the classroom?
    Can they be acquired and placed in an appropriate classroom habitat without harm to the animals?
    Does the acquisition pose any harm to the students?

Care and use
    Habitat
    Is the classroom habitat safe for the animal? Are temperature, humidity, appropriate?
    Is proposed classroom activity appropriate to activity cycle?
    Are materials appropriate for digging, nesting foraging, tunneling, etc.?
    Does the habitat provide appropriate options for movement, rest.

    Social life
    Is habitat appropriate to the type, frequency, intensity of social contact typical of this species?
    If there is more than one individual in an enclosure, how should they be matched or mixed by age, sex, size, or other important variables?
    Is there adequate opoortunity for access to food, water, hiding places for all individuals in a social group?

Life cycle needs
    Is there adequate opportunity for physical growth and development or social maturation?
    Can normal life cycle functions such as reproduction and birth/hatching be carried out?
    If reproduction is successful, can the offspring survive and thrive in the classroom habitat?
    How and up to what point will this population growth be sustained?

Disposition
    What will happen to the animals after the completion of the lesson(s)?
    Can they return to their natural habitat?
    Is any sort of preparation, training, or rehabilitation required before the animal can return to nature or its previous way of life? Is so, how will this be carried out and by whom?
    How will the disposition of the animal(s) be introduced to and discussed with the students?

searchers and research technicians are frequent contributors to the journal *Humane Innovations and Alternatives* (Petto *et al.*, 1992; Cohen & Block, 1991; O'Neill, 1987). In recent years some winners of the journal's annual recognition award have also been on the research staff in biomedical research facilities (Anon., 1992, 1993).

Furthermore, one may argue that confronting the animal subject of our learning 'face-to-face' can be the basis of a sensitizing process in which the students learn about the real needs of non-human animals and the animals' observable reactions to handling, care, and educational activities of various sorts. The presence of living animals in the classroom can be a valuable way to increase the appreciation of learners for the real animal and its experience of life. This is done, not in 'cook-book' laboratory experiments nor improvised classroom 'habitats' for creatures nabbed in the schoolyard, but in carefully researched and considered exploration of the needs and way of life of the animals that might come into the learners' educational experience.

Rather than desensitizing the students to the animals that will enrich their education, this process requires the students to confront the real needs that living animals have in their environments. Direct, personal interactions with living animals provide the best opportunity for the bonding and empathetic responding between student and non-human animal that is universally acknowledged from Davis and Balfour (1992) to Shapiro (1990, 1991) to Weatherill (1993) and Ascione (1992). Because there are real and observable consequences in such a situation for making poorly informed choices about learning activities, habitat construction, or even choice of appropriate animal subject, students and teachers must confront and accept the consequences of their actions through interactions with living animals. Davis and Balfour (1992) argue that these interactions also bring benefits to human scientific and educational activities.

Another consequence of educational animal use is the development of an industry that serves the needs of thousands of schools that will use animals in some way. This is an important issue in Hepner's (1994) examination of the role of animals in education. The sheer volume of animals that must be killed, skeletonized, and/or preserved in some form every year in North America alone would probably surprise most educators. It is not only a matter of volume, but a matter of the effect on our expectations of the educational experience with animals.

These effects are visible in several ways. First, there are the offerings in the companies' catalogues. Schools and teachers may think that their options are defined, or at least limited by what major supply companies offer. When asked why they have chosen a certain species for anatomical study, it is not uncommon for biology teachers to make reference to what was available from some catalogue or supply house.

Secondly, the large-scale national supply houses act to remove us from the conditions of life and death of the animals before they arrive in the classroom. Even for animals that are not killed for educational use, our ignorance of the quality of their previous lives allows us not to include such questions in our humane education repertoire.

Thirdly, there are financial incentives, of course, to both the supply companies and the school districts for keeping a large-scale, centralized facility with 'efficient' production practices. There is some efficiency of scale with respect to costs, profit margins, storage, and delivery. There are volume discounts for school districts, and there is a certain ease to 'one-stop' catalogue shopping.

If the process to acquire each of the animals supplied from these sources took an approach similar to the one we propose here, then the existence of large, centralized supply houses that kill and preserve millions of animals annually might be somewhat less worrisome. However, the sheer volume of this industry's output should be enough to make us reconsider how our choices to use animals in education relates to this phenomenon. The realities of the animal supply business must be a part of the process of choosing to use animals in education.

If there is a determined need for animals or animal tissues in the classroom, the humane educational process is enhanced by the explicit discussion by teachers and learners of the questions of source and supply. Is it better to use purpose-bred animals, or specimens from slaughterhouses, or body parts from hunters or taxidermists? And, what social, economic and moral implications does each of these choices have? How should, or could, we decide among them and on what basis?

The process of recognizing the social implications of animal use beyond the classroom adds another important dimension to humane education. The whole learning community makes an informed and conscious choice for specific learning activities in which at least one component is animal based. It is vitally important that the learning community take this discussion beyond the blanket prescription or proscription of animal use.

Animal-based education that is humane

As this discussion has suggested, animal-based education that includes discussions of the ethics of animal use is humane education because it fosters a humane attitude (Downie, 1993). As set out in the draft standards for science education (NRC, 1994), students do research on the needs of an animal before it gets into the classroom, and this research includes whether a classroom environment is suitable for this animal.

The process of considering these issues for animals in education has three stages. The first stage focuses on pedagogical issues and is generally the domain of the teacher. These issues are laid out in Table 12.1 and have to do with the objective of the lesson, learning evaluation, integration of the animal-based activities with other aspects of the curriculum, and the design and presentation of the materials. It is also important to consider here the stage of cognitive development of the learners, their needs for current and future knowledge about animals, the specificity of the proposed animal use to their curricular or developmental needs, and their ability to articulate the complex emotional and empathetic concerns that they may feel.

The second stage focuses on the impact on the animals themselves. These issues are laid out in Table 12.2. How is the animal acquired? What is the animal's experience of life, and how would its use in education affect this experience? How would its physical and environmental needs be met? In the longer term, what will be the effects of this use on the animal and what will its future be?

The third stage focuses on the wider social impact. What humane principles can we instil or support in the community of learners with this process? What are the effects of this process and its outcome (whether to use animals) on the students and the school? Does the process of making the choice help students to understand the wider social impacts of the biological supply industry?

Conclusions

What we have proposed here is an outline for making the choice to include animals in the curriculum a humane learning activity. All members of the learning community should be actively engaged in the process of constructing the humane ethic that will govern the choice to use animals in the classroom and the decisions on how they will be used. It must be clear from the start that there is a choice to be made. We wish to avoid the phenomenon described by McGinnis (1992) of beginning with the conclusion that animal use in education is automatically either 'noa' or 'taboo' – prescribed or forbidden. When the outcome of this consideration is not a foregone conclusion, the process of making these choices adds a valuable dimension to the educational process for all members of the learning community.

For the whole learning community, this process of considering the various practical issues of acquisition, classroom care and use, and disposition of animals used in educational activities is the essence of humane education, because it requires the students to confront these issues explicitly. In so doing, it shows that the teacher and the school value the animals as entities in themselves worthy of such consideration and not only as a means to an end.

In the end, taking this process seriously may mean, perhaps, that some activities using animals in the classroom will not be done at particular times and places – even when they clearly have pedagogical value. It may mean that there will be several learning activities and that not all students will participate in each of them. It may mean that the curricular activities involving animals will be developed around different choices. However, none of what we have described as the process of humane education means that these learning activities will never be done. In the end 'humane' education is a process that increases, not decreases sensitivity of all the members of the learning community to the impact of their learning. This, we believe, can be accomplished through a process of active consideration of these impacts in the various dimensions that are affected by these choices.

Acknowledgements

We are grateful for the critical comments and suggestions on earlier drafts of this manuscript from our colleagues Alan Beck, Hugh LaFollette, Laura McMahon, Lisa Wachtel, and Tami Wolden-Hansen. We would also like to acknowledge the contributions to the development of the ideas in this paper in the course of earlier (real and virtual) conversations with Arnold Arluke, Maria Boccia, Judith Golden, Harold Herzog, Dietrich von Haugwitz, Brian Luke, Tere Ma, Andrew Rowan, Peter Singer, Millard Susman, Gary Varner, and Lynn Willis. This collaboration was supported in part by a grant to the University of Wisconsin–Madison Center for Biology Education from the Howard Hughes Medical Institute, by the University of Wisconsin-Madison, and by NIH (NCRR) grant 000167 to the Wisconsin Regional Primate Research Center. This paper is WRPRC Publication number 37–041. We also acknowledge the generosity of Judith Schrier whose support initiated our collaboration on these issues with a gift in memory of her late husband Allan Schrier and their parents, Jean Schrier, Bernard Sanow, and Ruth Sanow.

References

Anon. (1992). PSYeta's *Human Innovations and Alternatives* Annual Award, 1992. Viktor Reinhardt. *Humane Innovations and Alternatives*, **6**, 317.

Anon. (1993). PSYeta's *Human Innovations and Alternatives* Annual Award, 1993. Peggy O'Neill Wagner. *Humane Innovations and Alternatives*, **7**, 423.

Arnheim, V. (1969). *Visual Thinking*. Berkeley: University of California Press.

Ascione, F. R. (1992). Enhancing children's attitudes about the humane treatment of animals: generalization to human-directed empathy. *Anthrozoös*, **5**, 176–91.

Ascione, F. R. (1993). Children who are cruel to animals: A review of research and implications for developmental psychopathology. *Anthrozoös*, **6**, 226–47.

Atran, S. (1990). *Cognitive Foundations of Natural History: Towards an Anthropology of Science*. New York: Cambridge University Press.

Beck, A. M. (1968). Behavior of dogs: canid behavior in a natural setting. In *Animal Behavior in Laboratory and Field*, ed. E. O. Price & A. W. Stokes, pp. 125–7. San Francisco: W. H. Freeman and Co.

Beck. A. M. (1980). The vertebrate animal in high school biology. In *Animals in Education: Use of Animals in High School Biology Classes and Science Fairs*. ed. H. McGiffin & N. Brownley, pp. 60–5. Washington, DC: Inst. for the Study of Animal Problems.

Benham, D. C. (1991). A short stay, a long-lasting lesson. *Science and Children*, **29**(3), 19–21.

Bisbee, G. D. (1990). Animal awareness. *The Science Teacher*. **57**(3), 31.

Brown, T. C. (1991). *Human Universals*. Philadelphia: Temple University Press.

Cantor, D. (1992). Animals don't belong in school. *American School Board Journal*, **179**(10), 39–40.

Cohen, P. S. & Block, M. (1991). Replacement of laboratory animals in an introductory-level psychology laboratory. *Humane Innovations and Alternatives*, **5**, 221–5.

Davis, H. & Balfour, D. (1992). *The Inevitable Bond: Examining Scientist–Animal Interactions*. New York: Cambridge University Press.

Donnelley, S. (1994). Exploring ethical landscapes. In *The Brave New World of Animal Biotechnology*. ed. S. Donnelley, C. R. McCarthy & R. Singleton, Jr., Hastings Center Report, 1994; Suppl. **24**(1), S3–4.

Donnelley, S. (with Dresser, R. , Kleinig, J. & Singleton, R.). (1990). Animals in science: the justification issue. In *Animals, Science, and Ethics*, ed. S. Donnelley & K. Nolan, Hastings Center Report, Suppl. **20**(3), 8–13.

Downie, R. (1993). The teaching of bioethics in the higher education of biologists. *Journal of Biological Education*, **27**(1), 34–8.

Downie, R. & Alexander, L. (1989). The use of animals in biology teaching in higher education. *Journal of Biological Education*, **23**(2), 103–11.

Driscoll, J. W. (1992). Attitudes toward animal use. *Anthrozoös*, **5**(1), 32–9.

Felthous, A. R. & Kellert, S. R. (1986). Violence against animals and people: Is aggression against living creatures generalized? *Bulletin of the American Academy of Psychiatry Law*, **14**, 55–69.

Field, P. & Shapiro, K. (1988). A new invasiveness scale. *Human Innovations and Alternatives*, **2**, 43–6.

Frey, R. (1987). *The World of the Crow Indian: As Driftwood Lodges*. Norman: University of Oklahoma Press.

Friedman, A. (1991). A big lesson in a small pond. *Science and Children*, **28**(5),

27–9.

Gardner, H. (1993). *Multiple Intelligences: The Theory in Practice*. New York: Basic Books.

Gillespie, A. K. & Mechling, J. (eds.) (1987). *American Wildlife in Symbol and Story*. University of Tennessee: Knoxville.

Hatano, G. , Siegler, R. F. , Richards, D. D. , Inagaki, K. , Stavy, R. & Wax, N. (1993). The development of biological knowledge: a multi-national study. *Cognitive Development*, **8**(1), 47–62.

Hepner, L. A. (1994). *Animals in Education: The Facts, Issues, and Implications*. Alberquerque, NM: Richmond Publishers.

Kale, M. (1993). A world of animals just beyond our vision. *Interactions*, **11**(1), 10–12.

Katcher, A. , Beck, A. M. & Levine, D. (1989). The PAL Projects at Lorton. *Anthrozoös*, **2**(3), 175–80.

Keiser, T. D. & Hamm, R. W. (1991). Forum: dissection: the case for. *The Science Teacher*, **58**(1), 13, 15.

Kellert, S. R. & Felthous, A. R. (1985). Childhood cruelty towards animals among criminals and noncriminals. *Human Relations*, **38**, 1113–29.

Kinzie, M. B. , Strauss, R. & Foss, J. (1993). The effects of interactive dissection simulation on the performance of high school biology students. *Journal of Research in Science Teaching*, **30**(8), 989–1000.

Jungck, J. R. (1985). A problem-posing approach to biology education. *The American Biology Teacher*, **47**(5), 264–6.

LaHart, D. E. (1991). Squirrels – a teaching resource in your schoolyard. *Nature Study*, **44**(4), 20–2.

Lévi-Strauss, C. (1965). *Le Totémisme Aujourd'hui*. Paris: Presses Universitaires de France.

Lock, R. (1993). Animals and the teaching of biology/science in secondary schools. *Journal of Biological Education*, **27**(2), 112–14.

Lock, R. (1994). Biology – the study of living things? *Journal of Biological Education*, **28**(2), 79–80.

Löfgren, O. (1985). Our friends in nature: class and animal symbolism. *Ethnos*, **50**(3–4), 184–213.

Lott, D. F. (1988). Feeding wild animals: The urge, the interaction, and the consequences. *Anthrozoös*, **2**(4), 255–7.

Lowenfeld, V. & Brittain, W. L. (1970). *Creative and Mental Growth*, 5th edn. New York: Macmillan.

Majerus, M. E. N. , Kearns, P. W. E. & Forge, H. (1989). Ladybirds as teaching aids: 2. Potential for practical and project work. *Journal of Biological Education*, **23**(3), 187–92.

Mansbach, R. S. & Simmonds, R. C. (1986). Compassionate teaching in the animal lab: an ethical approach to animal research must start early. *The Science Teacher*, **53**(1), 46–8.

Markova, D. & Powell, A. R. (1992). *How Your Child is Smart: A Life-changing Approach to Learning*. Berkeley, CA: Conari Press.

Mayer, V. I. & Hinton, N. K. (1990). Animals in the classroom: considering the options. *The Science Teacher*, **57**(3), 27–30.

McGinnis, J. R. (1992). The taboo and the 'noa' of teaching science-technology-society (STS): a constructivist approach to understanding the rules of conduct teachers live by. Paper presented at the annual meeting of the Southeastern Association for the Education of

Teachers of Science, Wakulla Springs FL. Feb 14–15.

Milgram, S. (1974). *Obedience to Authority*. NY: Harper and Row.

National Research Council, National Committee on Science Education Standards and Assessment. (1994). *National Science Education Standards*. Washington, DC: National Academy Press.

Offner, S. (1993). The importance of dissection in biology teaching. *The American Biology Teacher*, **55**(3), 147–9.

O'Neill, P. L. (1987). Enriching the lives of primates in captivity. *Humane Innovations and Alternatives*, **1**, 1–5.

Orlans, F. B. (1991). Forum: dissection: the case against. *The Science Teacher*, **58**(1), 12, 14.

Peterson, N. S. & Jungck, J. R. (1988). Problem posing, problem solving, and persuasion in biology education. *Academic Computing*, **2**(6), 14–17, 48–50.

Petto, A. J. , Russell, K. D. , Watson, L. M. & LaReau-Alves, M. L. (1992). Sheep in wolves' clothing: Promoting psychological well-being in a biomedical research facility. *Humane Innovations and Alternatives*, **6**, 366–70.

Petto, S. G. (1994). Time and time again: holistic learning through a multimodal approach to art history. MFA Thesis, Boston University.

Quentin-Baxter, M. & Dewhurst, D. (1992). An interactive computer-based alternative to performing rat dissection in the classroom. *Journal of Biological Education*, **26**(1), 27–33.

Regan, T. (1993). Ill-gotten gains. In *The Great Ape Project: Equality beyond Humanity*, ed. P. Cavalieri & P. Singer. New York: St. Martin's Press.

Rodd, R. M. (1990). *Biology, Ethics, and Animals*. Oxford: Oxford University Press.

Sapontzis, S. F. (1987). *Morals, Reason, and Animals*. Philadelphia: Temple University Press.

Shapiro, K. (1990). The pedagogy of learning and unlearning empathy. *Phenomenology and Pedagogy*, **8**, 43–8.

Shapiro, K. (1991). The psychology of dissection. *The Animals' Agenda*. pp. 20–1.

Texley, J. (1992). Doing without dissection. *The American School Board Journal*, **179**(1), 24–6.

Weatherill, A. (1993). Pets at school: Child–animal bond sparks learning and caring. *InterActions*, **11**(1), 7–9.

Weissenger, J. (1990). Right before your eyes. *Nature Study*, **44**(1), 16–19.

13

'Minding animals': the role of animals in children's mental development

M. PATRICIA HINDLEY

Introduction

There is a profound, inescapable need for animals that is in all people everywhere, an urgent requirement for which no substitute exists . . . . It is the peculiar way that animals are used in the growth and development of the human person in those most priceless qualities which we lump together as 'mind'. It is the role of animal images and forms in the shaping of personality, identity and social consciousness. Animals are among the first inhabitants of the mind's eye. They are basic to the development of speech and thought [and] are indispensable to our becoming human in the fullest sense.   *(Shepard, 1978, p. 2)*

Shepard's thesis is that the presence of animals is integral to the development of human mind. In western culture at least, all children use animal imagery in the development of consciousness because, historically, thought arose, in large part, from the interactions between people and animals. This intimate connection linking animals to the development of speech, rationality, emotion and consciousness, is not the same as thinking about animals. Rather, it is what Shepard calls 'minding animals' (1978, p. 249); that is, it is in the act and nature of thought, and it is essential in the development of self-identity and self-consciousness (1978).

It is not only the presence of animals that is important to human development, but also an understanding of the nature of animal consciousness itself (Griffin, 1992). Considerable comparative research has been done in this area, raising questions not only about the nature of animal mind and the development of human mind, but also about the ethical treatment of animals. Western civilization's long-standing and predominantly negative attitudes towards animals suggests either ignorance of, or a profound disregard for, the indispensable role animals play in the development of consciousness in young children. In exploring that role, this chapter will examine the significance of the presence of animals in the development of children's emotion and cognition. It will also explore animal consciousness with respect to its role in the development of human mind.

Attitudes to animals

Animals provide, in their very existence, the possibility to experience a profound and unique companionship. In his essay, 'Why Look at Animals?', John Berger argues that our systematic destruction of wildlife and its habitats has destroyed that companionship forever in two important ways: we have physically and culturally marginalized animals. This leads to a conflict because human language, stories, dreams and myths constantly recall animals. Instead of being acknowledged, 'animals of the mind' have been co-opted into certain categories, especially the categories of 'the spectacle', 'the exhibit', and 'the performer'. In zoos, circuses and marine parks, animals have been 'immunized to encounter'. Berger writes that 'the look between animals and man, which may have played a crucial role in the development of human society, has been extinguished forever' (Berger, 1980, p. 26).

Similarly, James Carey, a communication scholar, has suggested that human history is simultaneously the history of the transformation of nature, the environment and therefore of reality. He asserts: 'There is now, virtually, no reach of space, of the macroscopic or the microscopic, that has not been refigured by human action. Increasingly, what is left of nature is what we have deliberately left there' (Carey, 1989, p. 73). There is also concern about the roots of our environmental crisis and its implications for the belief that animals are indispensable in our lives (Shepard, 1978). Shepard, a human ecologist, writes that animals will only exist at some cost. Unless we are prepared to pay the price, animals will no longer be a part of our experience (1978).

The uncompromising and pessimistic messages of Berger, Carey and Shepard underscore our current environmental crisis. Rationality has alienated us from other species as well as from other human beings. Rationalization of the domination of other individuals creates a serious split between human behaviour and social experience (Laing, 1967). This division happens through a complex process of denial, repression and mystification, which results in the alienation of ourselves from our experiences (Laing, 1967). These processes are acts of 'denial through distancing' (Livingston, 1989, p. 11).

In western cultures it is not surprising that behaviours such as 'denial through distancing' occur in order to separate us from other non-human animals. The human relationship with nature has always been ambivalent. We are grounded in it and yet are not of it because we are the creators of human culture, and this creates ambiguities within our firsthand experience of the forces of nature (Noske, 1986). The blurring of boundaries between 'self' and 'other' finds expression in the way that human characteristics are ascribed to animals. On the one hand, humans portray strong anthropomorphic tendencies while at the same time they greatly exploit non-human animal life. For example, cuddly animal toys and Disney cartoons are in sharp contrast to animal skins for clothing and animal flesh for food. However, the closer humans become to animals, the harder it is to exploit the animals in the same way as before (Serpell, 1986). 'Denial through distancing' enables humans to deal with animals as 'other', as humans similarly will respond to other humans. For instance, in times of war, the foe will often be objectified as 'the other', who is also 'less than human'. Having these alternatives and opposing views of animals presents a difficult paradox to unravel.

Humans typically create distancing behaviours in order to condone our modern treatment of animals (Serpell, 1986). Farmers, nurturing and naming many of their domestic animals, must isolate themselves from the abattoirs. Hidden behind faceless, windowless buildings, factory farming conceals the large-scale impersonal processes of modern food production. Medical and scientific research on experimental animals is equally well hidden. Other evidence of deliberate concealment is found in language. For example, slaughter houses are called 'meat plants'; slaughtered animals, 'meat'; mink pelts are 'harvested' rather than flayed; predators slated for extermination are called 'pests' or 'vermin'. Fur coats and canned tuna are not connected with the processes of trapping foxes or netting dolphins. Serpell (1986) has suggested that humans cannot 'afford' to think of animals as 'kin' because of our strong emotional ties to them.

Anthropomorphism and empathy

Why do humans create such elaborate rituals, rationalizations and justifications for their treatment of animals? Underlying distancing methods and anthropomorphic tendencies is the essential human need for profound relationships with other species. Additionally, and more significantly, a fundamental link between those anthropomorphic tendencies and the human capacity for empathic relationships may be indicated. Anthropomorphism is especially critical for children in learning how to love, trust, and to be just and moral. Anthropomorphism may well underlie children's ability to establish and be taught empathy. Empathy, as an emotional response, demands a cognitive element in that one individual must project his or her own feelings in the place of another's, or vice versa.

In the introduction to her novel, *Buffalo Gals and Other Animal Presences*, Ursula Le Guin eloquently underscores the centrality of anthropomorphism in keeping us connected to, and part of, the incredible diversity of life on our Planet Earth. Le Guin asks why animals in children's books talk, why a prince after eating fish scales can understand the language of mice, why the tortoise calls to the hare, 'I'll race you', when we all know, including young children, that animals do not talk. Le Guin reminds us that in western Christianity when St Francis conversed with animals it was astonishing, whereas in the eastern religion of Buddhism, it was considered commonplace for the Buddha to be transformed into a jackal. Le Guin rails against the isolationism of the Church as 'that soul-fortress', and also against western civilization itself, which is described as building 'its morality by denying its foundation'. Le Guin writes:

By climbing up into his head and shutting out every voice but his own, 'Civilised Man' has gone deaf. He can't hear the wolf calling him brother – not Master, but brother. He can't hear the earth calling him child – not Father, but son. He hears only his own words making up the world .... Children babble, and have to be taught how to climb up into their heads and shut the doors of perception .... This is the myth of Civilisation, embodied in the monotheisms which assign soul to Man alone. *(Le Guin, 1987, p. 11)*

Le Guin states that the continuity and interdependence of all life forms is a 'lived fact' for children and those whom civilization calls 'primitive'. Through ritual, myth and fiction, it is made conscious. This continuity of existence without sentiment, cruelty or benevolence in itself, is a fundamental cornerstone for 'whatever morality may be built upon it' (Le Guin, 1987, p. 11). Most of us, except children, have lost this connectedness; this loss of ability to 'come into animal presence' may be our greatest tragedy. The title of Denise Levertov's poem succinctly draws together these ideas when Le Guin writes:

Perhaps it is only when the otherness, the difference, the space between us (in which both cruelty and love occur) is perceived as holy ground, as the sacred place, that we 'can come into animal presence'. *(Le Guin, 1987, p. 13)*

The Hassidic philosopher, Martin Buber, also writes about the concept of the non-human in human thought and understanding. His poetic description of his boyhood experience with his horse speaks of 'coming into animal presence' and of mutual recognition and acceptance as equals.

... what I experienced in touch with the animal was the Other, the immense otherness of the Other, which, however, did not remain strange like the otherness of the ox and the ram, but rather let me draw near and touch it.... [I] felt the life beneath my hand, it was as though the element of vitality itself bordered on my skin, something that was not I, was certainly not akin to me, palpably the other, not just another, really the Other itself and yet it let me approach, confided itself to me, placed itself elementally in the relation of a *Thou* and *Thou* with me.... I was approved.   *(Buber, 1965, p. 23)*

Development of children's empathy

Shepard (1978) argues that animal otherness is intimately tied to the development of the human mind. For instance, the psychoanalyst Harold Searles (1960) has examined the role of the natural environment in normal and abnormal development. He finds that the plant and animal environment is of central significance in the formation of children's emotional stability, personal identity and a continuous sense of experience. Although he focuses on the entire natural environment, he emphasizes the important role that is played by animals in children learning 'relatedness'. Children experience most animals as relatively non-threatening, less demanding and intrusive, and less complex than their human world. Animals provide a safe ground from which children can develop certain essential human capacities by first seeing them reflected in the behaviour of animals. For instance, significant behaviours might be displays of affection or discipline between mother and infant; a mother's defence of her young; or her seeming indifference at the time of weaning. Furthermore, the nature of animals, free from parental injunction and ambivalence, allows children to test themselves. It allows them to see themselves more clearly, to see in their relationships with animals their own capacities for gentleness, empathy, cruelty and indifference. Berger writes that 'animals supply examples of the mind as well as food for the body. They carry not only loads but also principles' (1971, p. 1043).

Shepard (1978), in writing about the principles that animals represent, refers to the role that animal images and forms play in the shaping of identity and social consciousness, and in the development of speech and thought. Animal participation in each stage of the growth of consciousness is facilitated, Shepard asserts, through our innate and indispensable capacities for anthropomorphism, sympathy and curiosity (1978). In western cultures at least, in the first decade of life, the child is instinctively and intensely drawn to animals as genuine, direct and unambiguous beings:

mouse is mouse, goose is goose (Shepard, 1978). By contrast, the human social world is a quagmire of innuendoes, emotional complexities, blurred intents and subtleties. Thus, the child, through a process of interaction, mimicry and observation, is able to embrace and absorb the animal world in a straightforward manner.

Middle childhood, ages 5 to 10 years, is described as a special period where children experience nature in some 'highly evocative way, producing in the child a sense of profound continuity' (Cobb, 1959, p. 124). It is a period of 'plasticity of perceptual response', reflected in the child's play as a spontaneous effort to transform himself/herself into some other being (Cobb, 1959, p. 126). Imitation of animals is second nature to children, essential for self-expression and escape through make-believe (Lonsdale, 1981). Dance is seen as the perfect vehicle for this expression, an exuberant form of play and an essential part of games, in which children are free to become whatever they desire: heron, tiger, wizard, wind, bear or butterfly. Predator/prey dances allow children to feel powerful, but not omnipotent; to feel scared, but not overwhelmed. Animal imitation in dance and game contains powerful teachings about behaviour; for example, witness the teachings, especially for children, in aboriginal dances (Lonsdale, 1981).

Shepard (1991), observed that the talent to imitate is one of children's most effective ways of adapting to the world. Through plays, fairytales and dreams, children relate to animals as the embodiment of feelings. Predator/prey games like 'Cat and Mouse' or 'Rabbit-in-the-Hole', and exchange of identity games like 'Piggyback Riding', match co-opted movements to feelings, thus providing one way in which emotions are introjected. In addition, fairytale characters embody children's fears and anxieties about powerlessness, abandonment and loss of love. Many of these fairytale characters are animals, often in disguise, ready to transform ugliness into beauty, stupidity into wisdom, defeat into victory, creating in children's minds the possibility of becoming bear or butterfly. Issues too painful to think about may present themselves again in animal disguise, especially in the dreams of young children (Shepard, 1978, 1983, 1991).

In longitudinal studies on children's dreams, animals are considered to be the predominant characters for children aged three to five (Foulkes, 1982). There is also a strong blurring of boundaries between a child's sense of self and the dream character. Similarly, Pitcher and Prelinger's (1960, quoted in Foulkes, 1982) data on children's stories reveal a preponderance of animal characters, as well as a fluidity between animals and the child's self. An animal, as a separate character at the beginning, may by the end have become 'child as animal' (Foulkes, 1982). Both Piaget and Freud have pointed to this fluid relationship between young children and animals (Foulkes, 1982). 'A child can see no difference between his own nature and that of animals; he is not astonished at animals thinking or talking in fairytales; ... Not until he is grown up does he become [so] far estranged from animals ... ' (Freud, 1917, quoted in Searles, 1960, p. 4). Nevertheless, there are problems in both these approaches. Neither author has explained why there is such a concentration of animal characters to the exclusion of human characters, even familial ones, in the dreams and daytime stories of young children.

Cognition

Foulkes' data for these young children (3–5 years) does not seem to support the Freudian notion of dreams as vehicles for unconscious desires and fears, or of dreams as carriers of deep feelings (1982). These children's dreams contained not familiar pets or exotic creatures, but rather the ordinary, un-named and non-threatening animals found in farmyards and back gardens, such as pigs, horses, birds and frogs (Foulkes, 1982). Dreams were most frequently simple; for instance, 'a bird singing', 'a frog in water', 'a horse running'. Foulkes (1982, 1993) and Cavallero and Foulkes (1993) suggest that, in young children, dreams are a reflection of their cognitive development. 'Dreaming is a much more "organized" process than is generally imagined and ... it employs the same systems of mental representation and mental processing as are exhibited in the waking phenomena' (Cavallero & Foulkes, 1993, p. 3). Meier (1993) found that dreams are not visual experiences as is commonly believed, but rather their content is about social interactions predominantly involving speech. In other words, dreaming is an intellectual rather than a perceptual process (Foulkes, 1993).

From the dreams of children in age groups 5–7 and 8–12 years, the self-reported evidence suggests a decrease in dream-character animals correlated with increased cognitive development (Foulkes, 1982). Animals are almost equally predominant in the 5–7 year age group, and then gradually taper off towards the end of middle childhood. Similarly, there is a gradual increase in the presence of self and other human characters in dreams. As children get older, there is marked evidence of the representation in dreams of a freely acting self-character who experiences and expresses emotions and who is seen as the dreamer. In a more recent study, it was proposed that only with the onset in development of a sense of self (at around 8 years of age, according to Piaget), is such complex dreaming possible (Foulkes, 1993). After age 8, it is only in the dreams of psychologically or cognitively impaired children that the cognitively simple and animal-dominated dreams remain (1993).

Further support for the theory that there is a direct link between a child's cognitive and social development and the presence of animals comes from evidence on dreams, play and stories, and indicates that young children are attracted to animals as sources of identification and mimesis (Fernandez, 1972; Bettelheim, 1976; Foulkes, 1982; Lonsdale, 1981; Shepard, 1991). Earlier researchers (Nelson, 1973; Bellak & Adelman, 1960, quoted in Foulkes, 1982), have speculated on what it is about animals which attracts children. They suggest that it is primarily an attraction to what is immediately eye-catching, namely animal movement (Nelson, 1973), and possibly to the impulsiveness of those movements (Bellak & Adelman, 1960), that results in such a preponderance of animals in children's earliest verbal language. What this means is that it is through animal action rather than through character that children identify with animals. Therefore, it is reasonable to assume that one essential purpose for this strong attraction is the important role played by animals in the mental growth of children. It would seem that young children, without their own ability to represent themselves effectively, need animals as external self-models. Foulkes (1982) speculates that it may well be that children's animal dreams are not creative or fanciful, but rather a 'compensation' for the inability to yet use the dream for self-representation: 'if animals are the earliest reliable extraself dream characters in a manifest-content sense, it should be understood that these seeming animal-others are, in fact, apprehensions of the still highly egocentric child herself or himself' (Foulkes, 1982, p. 82).

In cognitive development, the naming of animals is inseparable from speech development, in that naming in general prefigures categorical thinking (Shepard, 1991). This link involves more than simply thinking about animals: the 'connection is in the act and nature of thought, the working of mind' (Shepard, 1978, p. 2). The process of repetitive naming, so characteristic of young children, gradually includes what animals do. That is, names become verbs: to badger, to rat and to duck. In mimicking them, the child is not only like rabbit, but is rabbit. How animals move through the very fabric of a young child's physical, cognitive and emotional development from simple representation to complex symbolism is what Shepard refers to as the 'minding' or the 'swallowing' of animals (Shepard, 1991, p. 6): porpoise, weasel, crow; to be porpoise, to be weasel, to be crow; porpoising, weaseling, crowing; the 'friendliness' of the porpoise, the 'slyness' of the weasel, the 'cleverness' of the crow. 'A winged moment of joy, a scuttling shadow of fear or a crawling anxiety': these are all part of the 'mysterious world of the self as bestiary' (Shepard, 1991, p. 8).

There is a distinct significance in possessing a rich set of animal images for children's early cognitive and emotional development. These images have important implications for the ontogeny of thinking and our capacity to relate to others. They evoke experiences in childhood which remain with us throughout our adult lives. In turn, they provide an essential ground which enables us in adolescence to move towards mature human relationships, and in adulthood to understand more fully humans' intimate kinship with other animal life (Shepard, 1978).

Human and animal consciousness

The case made above is that the presence of animals plays a crucial role in the development of human consciousness. In addition, by understanding the nature of animal consciousness, more light can be shed upon early childhood development and on the emergence of consciousness. The exact nature, and even the concept, of animal consciousness is very controversial; it is that nature that is discussed below.

For at least the last seven centuries (in the western world), animals have been considered to be machines and the wilderness to be wasteland. In addition, with the advent of Behaviourism, it was postulated that some animals possess some mental processes similar to those of humans, but do not possess consciousness. That is, they do not have the awareness of the state of their own minds, nor any purposeful intention to action. However, according to Donald Griffin, 'mind' can be defined as having the disposition to behave, and the capacity to think consciously about objects and events (1992). This concept has been typically denied to animals (Griffin, 1992). Nevertheless, there have always been dissident voices in this long-standing and often acrimonious debate on the nature of the animal mind and on the possibility of animal consciousness, cognition and emotion. At the end of the 20th century, some experimental work in zoology, psychology, anthropology and ethology seems to indicate that animals have consciousness, and that some animals are capable of self-recognition, intentional behaviour and language comprehension (Pepperberg, 1987, 1993; Byrne & Whiten, 1988; Cheney & Seyfarth, 1991a,b; Herman, et al., 1993; Savage-Rumbaugh, 1993; Savage-Rumbaugh et al., 1993; Parker, Mitchell & Boccia, 1994; Patterson & Cohn, 1994; Savage-Rumbaugh & Lewin, 1994).

However, while such experiments suggest that we can no longer insist with absolute certainty that animals do not think, at the same time, they underscore the great challenge researchers face in even defining the above concepts, let alone in creating models that would produce definitive data. For example, Parker and Mitchell comment on the diverse and frequently contradictory research on self-awareness. Attempts to define it, to understand its origins and its adaptive significance have resulted in only 'a rich and heterogeneous array of hunches and hypotheses about the possible mechanisms of self awareness, and their ontogeny and evolution' (1994, p. 413). Nevertheless, the theoretical gap between what is defined as human and animal is narrowing. For instance, research in language studies with apes, birds and cetaceans has moved in two important directions (Pepperberg, 1987, 1991, 1993; Griffin, 1992; Herman *et al.*, 1993; Savage-Rumbaugh, 1993; Savage-Rumbaugh *et al.*, 1993; Savage-Rumbaugh & Lewin, 1994). First, in experimental research, there has been a deliberate shifting of focus from the evaluation of animal performance within a human context to a focus which has ecological validity for a particular species. This shift puts any human/animal comparative work on a more equitable and realistic footing. Second, as the research itself has become more creative and ingenious, scientists have been able to turn their attention to the animal's capacity for versatile behaviour (Byrne & Whiten, 1988). In other words, they found a way to examine an animal's flexible responses to a wide variety of stimuli rather than being restricted to innate or associative responses to stimuli. There is now some evidence in support of the idea that animals possess abilities for thought, awareness and complex cognition. This evidence comes from an examination of such behaviours as deception, intention and pre-meditated planning (Griffin, 1982, 1984, 1992; Mitchell & Thompson, 1986; Byrne & Whiten, 1988; Whiten & Byrne, 1988; Ristau, 1991; Herman *et al.*, 1993; Pepperberg, 1993; Savage-Rumbaugh, 1993; Savage-Rumbaugh *et al.*, 1993; Parker *et al.*, 1994; Savage-Rumbaugh & Lewin, 1994).

As the theoretical gap narrows, the debate about animal consciousness and self-awareness intensifies. Although, as indicated above, animals will never be human, many scientists believe the study of animal mind to be central to the understanding of human mind and its origins. In contrast, others, like Latto, claim that the subjective life of animals can never be known; therefore questions on that topic should not even be addressed. Any indirect methods of investigating animal minds has proven fruitless. 'Conscious awareness in other animals is a closed world about which we can do no more than speculate' (Latto, quoted in Radner & Radner, 1989, p. 192). Chomsky sees no point in comparative linguistic research on apes and children on the grounds that the communication systems of the two species are poles apart, making comparison meaningless (Chomsky, quoted in Radner & Radner, 1989, p. 170). There is a lot at stake in this debate. To accept the fact that animals are sentient beings is to raise some fundamental ethical issues which would certainly highlight the relationship of human consciousness to animal consciousness. Currently, there seems to be little chance of any major shift in attitude with respect to animal consciousness. It is true to say that humans will be the ones who have the most to lose by disregarding animals in this way.

Conclusions

Most of the recent work done in the field of human/animal relationships has focused on the role of companion animals in human health (Katcher & Beck, 1983; Anderson, Hart & Hart, 1984). Also, as has been mentioned, considerable comparative research has been conducted on the nature of animal mind, and this continues to challenge our treatment of animals. However, there seems to have been little similar sustained experimental work on children's dependency on animals for early cognitive and emotional development. Foulkes' longitudinal studies of children's dreams sheds some light on the role of animals, but it is not the central thrust of his work. Surely, what is urgently called for, at the very least, is a strong interdisciplinary research focus in this area.

For many, the ideas expressed in this paper may be unacceptable. The paper refers to the work of both contemporary and earlier researchers. Some of the traditions in which these earlier researchers were working may have been absorbed or discarded. Nevertheless, it is worth noting that important questions have been raised by those researchers, and more recent ones, both about the crucial role played by animals in the development of children's emotion and cognition and the nature of animal minds. Mind, as it has been defined, has always been seen as the unique human quality which gives us our distinctive humanity and separates us from animals. It is one thing to raise the issue of animal minds and their boundaries, it is quite another to suggest that the development of human mind is, in any way, intimately and profoundly linked to animal survival. If both statements are correct, namely that animals are sentient and that humans are psychologically dependent upon the existence of animals, then the implications for us ethically, economically and ecologically are enormous. We are caught in a bind here. The record shows that we are not prepared to reverse the wanton destruction of other species or to recognize that they are thinking animals. At the same time, we can hardly ignore the fact that animals may be of critical importance to our own mental development. If that theory can be supported, then it may be that, in its inherent selfish implications, it is the only one powerful enough to provoke us to save animals.

References

Anderson, R. K. , Hart, B. L. & Hart, L. A. (eds.) (1984). *The Pet Connection: Its Influence on our Health and Quality of Life*. Minnesota: Center to Study Human–Animal Relationships and Environments, University of Minnesota.

Berger, J. (1971). Animal world. *New Society*, 25th Nov. , p. 1043.

Berger, J. (1980). *About Looking. Why Look at Animals?* pp. 1–26. New York: Pantheon Books.

Bettelheim, B. (1976). *The Uses of Enchantment*. New York: Knopf.

Buber, M. (1965). *Between Man and Man*. New York: Collier Books.

Byrne, R. & Whiten, A. (eds.) (1988). *Machiavellian Intelligence*. Oxford: Clarendon Press.

Carey, J. (1989). *Communication as Culture*. Boston: Unwin Hyman.

Cheney, D. L. & Seyfarth, R. M. (1991*a*). Reading minds or reading behaviour? Tests for a theory of mind in monkeys. In *Natural Theories of Mind*, ed. A. Whiten, pp. 175–94. Oxford: Basil Blackwell.

Cheney, D. L. & Seyfarth, R. M. (1991*b*). Truth and deception in animal communication. In *Cognitive Ethology: The Minds of Other Animals*, ed. C. A. Ristau, pp. 127–51. New Jersey: Lawrence Erlbaum & Associates.

Cobb, E. (1959). The ecology of imagination in childhood. *Landscape*, **9**(1), 122–32.

Fernandez, J. (1972). Persuasions and performances: the beast in every body and the metaphors in everyman. *Daedelus*, 101(1).

Foulkes, D. (1982). *Children's Dreams: Longitudinal Studies*. New York: Wiley.

Foulkes, D. (1993). Children's dreaming. In *Dreaming as Cognition*, ed. C. Cavallero & D. Foulkes, pp. 114–32. London: Wheatsheaf.

Cavallero, C. & Foulkes, D. (eds) (1993). *Dreaming as Cognition* (introduction), pp. 1–17. London: Wheatsheaf.

Griffin, D. R. (ed.) (1982). *Animal Mind–Human Mind*. New York: Springer-Verlag.

Griffin, D. R. (1984). *Animal Thinking*. Cambridge: Harvard University Press.

Griffin, D. R. (1992). *Animal Minds*. Chicago: University of Chicago Press.

Herman, L. M. , Pack, A. A. & Morrel-Samuels, P. (1993). Representational and conceptual skills of dolphins. In *Language and Communication: Comparative Perspectives*, ed. H. L. Roitblat, L. M. Herman & P. E. Nachtigall, pp. 403–42. New Jersey: Lawrence Erlbaum & Associates.

Katcher, A. H. & Beck, A. M. (eds.) (1983). *New Perspectives on our Lives with Companion Animals*. Philadelphia: University of Pennsylvania Press.

Laing, R. D. (1967). *The Politics of Experience*. Harmondsworth: Penguin.

Le Guin, U. (1987). *Buffalo Gals and other Animal Presences*. New York: New American Library.

Livingston, J. (1989). The ecological imperative. *The Institute for the Humanities Newsletter, Simon Fraser University*, **3**(1), 6–11.

Lonsdale, S. (1981). *Animals and the Origin of Dance.* London: Thames and Hudson.

Meier, B. (1993). Speech and thinking in dreams. In *Dreaming as Cognition,* ed. C. Cavallero & D. Foulkes, London: Wheatsheaf.

Mitchell, R. W. & Thompson, N. S. (1986). *Deception: Perspectives on Human and Non-Human Deceit.* Albany: State University of New York Press.

Noske, B. (1986). *Humans and Other Animals: Beyond the Boundaries of Anthropology.* London: Pluto Press.

Parker, S. T. & Mitchell, R. W. (1994). Evolving self-awareness. In *Self-Awareness in Animals and Humans,* ed. S. T. Parker, R. W. Mitchell & M. L. Boccia, pp. 413–28. New York: Cambridge University Press.

Parker, S. T. , Mitchell, R. W. & Boccia, M. L. (eds.) (1994). *Self-awareness in Animals and Humans.* New York: Cambridge University Press.

Patterson, F. G. P. & Cohn, R. H. (1994). Self recognition and self-awareness in lowland gorillas. In *Self-Awareness in Animals and Humans,* ed. S. T. Parker, R. W. Mitchell & M. L. Boccia, pp. 273–290. New York: Cambridge University Press.

Pepperberg, I. (1987). Interspecies communication: a tool for assessing conceptual abilities in the African grey parrot. In *Cognition, Language and Consciousness: Integrative Levels,* ed. G. Greenberg & E. Tobach, pp. 31–56. New Jersey: Lawrence Erlbaum & Associates.

Pepperberg, I. (1991). A comparative approach to animal cognition: a study of conceptual abilities of an African grey parrot. In *Cognitive Ethology: The Minds of Other Animals,* ed. C. Ristau, pp. 153–86. New Jersey: Lawrence Erlbaum & Associates.

Pepperberg, I. (1993). Cognition and communication in an African grey parrot (*Psittacus erithacus*): studies on a nonhuman, nonprimate, nonmammalian subject. In *Cognition and Communication: Comparative Perspective,* ed. H. L. Roitblat, L. M. Herman & P. E. Nachtigall, pp. 221–48. New Jersey: Lawrence Erlbaum Associates.

Radner, D. & Radner, M. (1989). *Animal Consciousness,* New York: Prometheus Books.

Ristau, C. (ed.) (1991). *Cognitive Ethology: The Minds of Other Animals.* New Jersey: Lawrence Erlbaum & Associates.

Savage-Rumbaugh, E. S. (1993). Language learnability in man, ape and dolphin. In *Cognition and Communication: Comparative Perspectives,* ed. H. L. Roitblat, L. M. Herman & P. E. Nachtigall, pp. 457–84. New Jersey: Lawrence Erlbaum & Associates.

Savage-Rumbaugh, S. & Lewin, R. (1994). *Kanzi: the Ape at the Brink of the Human Mind.* New York: John Wiley.

Savage-Rumbaugh, E. S. , Murphy, J. , Sevcik, R. A. , Brakke, K. E. , Williams, S. L. & Rumbaugh, D. M. (1993). *Language Comprehension in Ape and Child.* Monographs for the Society for Research in Child Development. Serial No. 233, 58(3–4).

Searles, H. F. (1960). *The Nonhuman Environment in Normal Development and in Schizophrenia.* New York: International Universities Press.

Serpell, J. (1986). *In the Company of Animals: A Study in Human–Animal Relationships.* New York: Basil Blackwell.

Shepard, P. (1978). *Thinking Animals: Animals and the Development of Human Intelligence.* New York: Viking Press.

Shepard, P. (1983). The ark of the mind. *Parabola,* **11**(2).

Shepard, P. (1991). Animals and identity formation. *Journal of Contemporary Thought*, 1.

Whiten, A. & Byrne, R. W. (1988). Tactical deception in primates. *Brain and Behavioural Sciences*, **11**, 233–73.

14

Alternatives to using animals in education

DAVID DEWHURST

Introduction

The use of animals to demonstrate known facts and/or principles still features in many biology courses in secondary schools and, perhaps more significantly in universities, despite a growing ethical concern amongst students and teachers. The morality of using animals in this way is controversial and the issues have been debated by several authors (Orlans, 1988; Langley, 1991; Smith, 1992). Since the introduction of the Animals (Scientific Procedures) Act 1986 in Britain (HMSO, 1987) experiments using live animals in schools have been outlawed, although such experiments are still legal in universities. In the university sector most animals are used in laboratory practicals, designed to supplement formal taught sessions, where they demonstrate known facts or principles. They are also used in undergraduate research projects. The number of animals used in teaching is significant (9164 procedures under Home Office regulations in 1992), although small when compared with the total number of animal procedures (2 854 046 in 1992) which include those performed in drug and cosmetic testing, medical research, etc. (see Table 14.1).

It might be expected that the high media profile of animal rights issues in recent years, which has heightened the awareness of both teachers and students to the unnecessary use of animals, would have led to a marked reduction in animal experimentation for teaching purposes. In addition, the large increase in the cost of animals (Hughes, 1984) and the decrease in unit funding in higher education would suggest that the use of animal experimentation, at least for teaching purposes, should be decreasing rather than increasing. This trend is evident in The Netherlands (Nab, 1990). The Home Office figures do not show this change occurring in the United Kingdom. The figures suggest that the total number of animals used is declining in the UK, while it would appear that the number used for

Table 14.1. *Total numbers of animal procedures and number of procedures categorized as education 1987–1992*

Year	1987	1988	1989	1990	1991	1992
Total number of procedures ($\times 1000$)	3631.4	3480.3	3315.1	3207.1	3164.6	2854.1
Number of procedures in teaching ($\times 1000$)	8.7	8.2	12.1	10.4	12.0	9.2

(Statistics of Scientific Procedures on Living Animals in Great Britain 1987; 1988; 1989; 1990; 1991; 1992: HMSO, UK)

education increased from 1987 to 1991 and even though the figure fell in 1992 it is still above that in 1988 (Table 14.1). At least part of this increase may be explained by the change in the regulations in 1986 which required the inclusion of procedures which were not covered by the previous Act, such as pithing of amphibia prior to experimentation. In most instances the number of animals used equates to the number of procedures, although it should be recognized that, in some circumstances, several procedures may be carried out on the same animal. Equally, the statistics do not include animals which have been freshly killed and tissue removed to be used in class practicals, providing euthanasia was performed by a Schedule 1 method of the Act.

While moral concerns, as evidenced by the rise in the numbers of students boycotting practical classes (MacDonald & Saano, 1987), and financial constraints may stimulate a demand for a change in teaching methods, this will only come about if realistic alternatives are available, which can be shown to be cost-effective and easier to use than traditional methods. In some cases, it may also be necessary for examination boards to change their requirements. For example, until recently, most of the examination boards in the UK included animal dissection as part of the syllabus for all students studying biology at A-level, even though a significant proportion would not pursue further studies, or a career, in biology.

Animal dissection

Clearly, a knowledge of anatomy is essential to understanding physiological function, but it is questionable whether dissection is the most appropriate way to achieve this knowledge. Indeed, one evaluative study showed that students at the Ohio State University learn anatomy as effectively from interactive videos as they do from cadaver dissection (Guy & Frisby, 1992). Barker (1988), in another study, found lecture-demonstrations to be as effective. The requirement for students to undertake animal dissection in high school courses has now been dropped, largely due to pressure from animal rights groups, and, since the early 1990s, no school has been obliged to dissect animals as part of the British National Curriculum. Despite this, many schools continue to dissect and a recent survey in one urban education authority found that in excess of 30% of secondary school teachers are still using dissection in lessons (Croall, 1994) even though a study of 14- and 15-year-olds found that many claimed to have learned nothing from class dissections (Lock, 1993; Lock & Millett, 1992).

In higher education a number of universities have responded to student unease about vivisection by offering opt-outs at the start of the course. In one university, where first-year biology students were given a choice between dissecting a rat or opting out and doing an equivalent laboratory exercise using a model, approximately 10% chose to opt out over a 5-year period (Downie & Meadows, 1995). Interestingly, this study also demonstrated that the opt-out students achieved similar examination results to those who performed the dissection. In other universities, the response has varied considerably: some have virtually abandoned using animals for teaching, while others insist that students sign a declaration saying they will take part in animal experiments before they are admitted on to the course.

Alternatives to animal dissection are widely available. Models of anatomical structures may prove useful although they often lack realism. A silicon model of a rat, with replaceable tails, has recently been developed at the Keio University in Tokyo in conjunction with the Koken Company of Japan. This is designed to be used primarily to allow students to practice administration of substances either intravenously by injection into a tail vein, or by oral routes.

Videotapes showing anatomical structures and dissection technique are useful teaching aids. However, the lack of interaction and the linear presentation of material makes them unsuitable alternatives in their own right. More interactive alternatives are also available. For instance, computer programs such as 'The Rat Stack' (Quentin-Baxter & Dewhurst, 1992), 'Operation Frog' (Scholastics Software, Scholastic Inc., 730 Broadway, New York, NY 10003; 1984) and interactive videodisks such as 'The Rat Anatomy Disk' (Quentin-Baxter & Dewhurst, 1995). 'The Rat Stack ' is a Hypercard stack running on Macintosh microcomputers, which allows students to interact with computer-generated images of a rat in various stages of dissection. These stages are presented to simulate the removal of superficial layers to reveal the structures beneath. Photographic images and line drawings also show important organs and tissues (down to electron microscope level in some cases), and combine with hypertext and features such as animation to make an effective independent learning package to teach the functional anatomy of the rat. 'The Rat Anatomy Disk' takes this idea a stage further and provides students with a realistic alternative to performing a mammalian dissection. The material covers in detail five areas of a rat dissection, (external features, abdomen, thorax, neck and head), and a large number of histology sections which illustrate the cellular detail of many mammalian tissues. Text and images presented on a computer screen, and organized as a Hypercard stack, are supported by high-quality, full-screen, full-colour moving video sequences on a separate video monitor. The dissection images are supported by a large amount of on-line information which is accessed in a highly interactive way.

Animals used in undergraduate biology laboratory classes

Laboratory-based classes involving animal experimentation feature prominently in many undergraduate courses in biological, medical, dental and health sciences particularly where physiology, biochemistry or pharmacology are major components. It is questionable, on moral, educational and economic grounds, whether, for many students, the continued use of large numbers of animals is justified. Animal experimentation is frequently used simply to demonstrate known facts or principles and for many students, particularly those who will pursue a career outside research, there is little educational justification for the continuation of such classes. Teachers, for many years, have been able to justify this traditional approach since no alternative approaches were available. However, the increased availability of microcomputers, with graphics capability, at all levels of education has led to the development of a wide range of potential alternatives. It is important that any proposed alternative satisfies a number of criteria which are as follows:

(a) capable of achieving primary teaching objectives and not disadvantage students;
(b) interactive;
(c) easy to use; and
(d) cost-effective.

A number of possible alternatives exist. For example, in biochemistry, practical class experiments may be redesigned to use plant rather than animal tissue or *in vitro* cultured cell lines. Brownleader (personal communication) has developed alternative practical (laboratory) schedules for a number of standard biochemistry practicals in response to letters from students disturbed by the use of animals in undergraduate courses. For instance, a laboratory class common to many undergraduate biochemistry courses uses rat or guinea pig liver as a source of mitochondria. Brownleader suggests that mitochondria may be isolated very simply from plant material such as beet, cauliflower or potato (tissues which avoid contamination with chloroplasts) and used successfully to study the effect of the respiratory chain inhibitors rotenone and antimycin A, and the uncoupler 2,4-DNP upon oxidative phosphorylation. The techniques required are similar to those used for isolation of the organelle from liver so that students still have the opportunity to practice laboratory skills. Similarly, he argues, that experiments which use rat brain tissue to study the effect of cofactors on glycolysis and the determination of lactic acid, may be replaced by experiments on plant tissue (e.g. red beet, *Beta vulgaris L.*). Again, the same isolation techniques are required as for animal tissue and all of the teaching objectives are fulfilled. Clearly, where realistic alternatives exist and the experimental protocol has been well developed there is no excuse for the continued use of animal tissue.

In physiology and pharmacology students will undertake practical classes in which they act as subjects. Thus they may measure heart rate, blood pressure, respiration and lung function, sensation, effects of exercise, nervous reflexes, etc. These may be seen as alternatives to using animals and in many universities and schools the number of practical classes involving human subjects has increased and the number of practicals using animals has decreased. However, in the majority of undergraduate courses animals are still used. Most experiments are performed on isolated tissue taken from freshly killed animals. For example, the sciatic nerve of the frog to demonstrate the properties of nerve; isolated segments of guinea-pig ileum are used to demonstrate the action of various pharmacological agents; and the isolated perfused rabbit heart is used to demonstrate the effects of sympatho- and parasympathomimetic agents on the heart. In most cases the tissue responses, for example, to electrical stimulation or the administration of a test substance, are displayed either on an oscilloscope or chart recorder. One possible alternative to such experiments is the 'Bio Videograph' (Harvard Apparatus Ltd.). This will play back pre-recorded videotapes containing sound, vision and analogue data recorded from real experiments. The recordings are played back with the synchronized analogue data output to any number of analogue chart recorders. Thus, for example, students could watch a recording of an animal experiment on closed circuit TV with normal sound commentary and monitor, on instruments distributed throughout the laboratory, the animals' blood pressure and respiration rate in response to injected drugs. A range of videos are available, many depicting common animal experiments. Typically, the investigator is shown preparing the animal or animal tissue and then performing a series of experiments, e.g. administering a drug or drug combination to the preparation. The tissue response is simultaneously displayed on those chart recorders linked to the BioVideograph. Students attend a linked chart recorder and are instructed via the video film when to depress appropriate buttons on the recorder. In this way they receive a copy of the results shown on the video film and can then take this printed data away to mull over at their leisure and use it to produce a report. This approach is used in a number of universities although to equip a laboratory to enable several students to use the instrument would be expensive. The level of interaction is low and the time taken to standardize the chart recorders and collect the data may be considerable.

Microcomputer simulations of animal experiments

A second alternative is to use microcomputers which are increasingly available in science departments at relatively low cost. In addition to running standard word-processing, spreadsheet, database, and statistical packages, microcomputers may be used to capture and display experimental data in a form not dissimilar to the real situation (i.e. function as a simulated oscilloscope or chart recorder; Webster and Myers, 1986; Molloy, Newport & White, 1986; Basarab-Horwath *et al.*, 1986). This data, if stored on disc, may then be used as the basis for the development of interactive computer programs which simulate experiments and offer a realistic alternative to using animals. A number of such programs have been described, which simulate animal experiments and use actual data to generate realistic-looking, simulated tissue responses (e.g. Brown *et al.*, 1988; Dewhurst, Brown & Meehan, 1988*a,b*; Dewhurst *et al.*, 1994; Dewhurst & Howells, 1990; Dewhurst & Ullyott, 1991; Clarke, 1987, 1988). An alternative strategy is to develop software which simulates an animal preparation and makes use of mathematical models or algorithms to generate simulated responses (e.g. Brown, Dewhurst & Williams, 1994; Hughes, 1984, 1987; Dewhurst & Meehan, 1989; Dewhurst & Williams, 1993; Dewhurst *et al.*, 1992).

Interactive software has been designed to offer a highly flexible teaching alternative, illustrating known principles to students in physiological and pharmacological laboratory practicals. These user-friendly programs are available for a variety of hardware platforms (e.g. IBM compatible, Macintosh, Nimbus PC 186, and BBC computers). The programs are typically menu driven and make use of an easy-to-use, Windows-like menu display from which a number of options may be selected (e.g. a menu option called HELP gives on-screen instructions on how to use the program). Most programs have INTRODUCTION and METHODS sections which present details of the preparation, the apparatus and the experimental·method using text and high-resolution colour graphics.

The use of actual experimental data allows realistic-looking simulated responses to be generated. The biological variation seen in the real experiment, from which the data were derived, is precisely reproduced in the interactive software programs. These programs provide great variation in the types of experiments that can be simulated, for instance, currently available programs include: Nerve Physiology, Muscle Physiology, Frog Heart, Neuromuscular Pharmacology, Cat Nictitating Membrane, Rabbit (Langendorff) Heart, Intestinal (Colonic) Motility, Respiratory Pharmacology, The Finkleman Preparation and Inflammation Pharmacology. Many of the programs are also supported by printed learning material in the form of comprehensive tutor's notes with suggestions on how to make full use of the program and a student-centred workbook that contains results tables and questions/assignments relating to the experiments.

Features such as animation are used where possible to enhance presentation. For example, in the introduction to the Nictitating Membrane program the basic pharmacology of neurotransmission at both the ganglionic synapse and the neuroeffector junction is reviewed. Here, information is presented as text alongside a high-resolution graphic representation of the synapse. As the user scrolls through the text describing the basic pharmacology of the synapse, the graphical image becomes animated and a picture which illustrates the possible sites of action of various drug types is gradually constructed (Fig. 14.1). This information may then be used in the EXPERIMENTS section of the program to assist students to design suitable investigative experiments.

The main section in all programs is EXPERIMENTS from which the user may simulate the performance of a number of experiments (e.g. demonstrating the investigation of the action of a particular drug on the preparation). High resolution colour graphics are used to produce realistic-looking (tissue) responses on the monitor which are displayed in a form

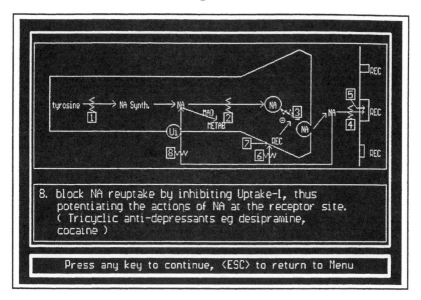

Fig. 14.1 Screen display from the INTRODUCTION of the 'Cat Nictitating Membrane' program showing a graphical representation of the neuroeffector junction. As the user scrolls through the text in the text window the graphic display is gradually built up and the potential sites of action of different drug types is illustrated.

comparable to the actual experiment (e.g. a storage oscilloscope or a chart recorder in which the responses are scrolled across the screen). Here, the level of user-interaction is high: students have a certain amount of control over how the experiment is conducted and are expected to collect data in much the same way as they would in the real experiment. Thus students may either progress through a series of exercises which allow them to collect data from pre-defined experiments, or take an investigative approach to a problem and design experiments (i.e. choose a drug or drug combination, for example, a drug plus an antagonist or potentiator), select the dose, route and site of administration, and determine stimulus parameters (stimulus voltage, frequency, duration).

These programs allow students to work through a series of tasks typical of those they would meet if they performed the *in vivo* experiment. For example, in the 'Nerve Physiology' program the computer simulation uses data obtained from actual experiments to generate high-resolution graphic simulations of compound nerve action potentials (CNAPs) which are displayed on the monitor in a form comparable to a storage oscilloscope (Fig. 14.2). Each of the available experiments has one variable (e.g. stimu-

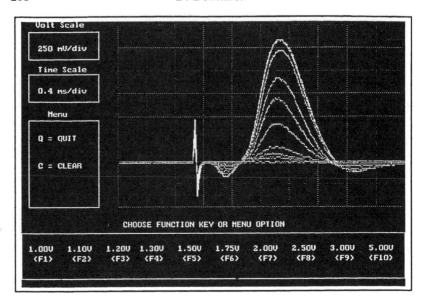

Fig. 14.2 Screen display from Nerve Physiology showing simulated compound nerve action potentials presented on a simulated oscilloscope screen. In this experiment pressing a function key (F0–F10) retrieves a file from disk (created from actual experimental data) appropriate to a particular stimulus voltage. Students take measurements directly from the monitor in much the same way as they would in the live experiment.

lus voltage, stimulus separation of paired stimuli, and distance between recording and stimulating electrodes) which is controlled from the keyboard. Thus, in an experiment to investigate the relationship between stimulus voltage and CNAP amplitude, the stimulus voltage may be selected from a range extending from 1.0 V (subthreshold) to 5.0 V (supramaximal). Selection of a certain stimulus voltage retrieves a file from disc appropriate to that stimulus and the simulated response is displayed (Fig. 14.2). The stimulus voltages have been selected to illustrate the important points (i.e. the range of thresholds of neurones within the sciatic nerve), and students can determine the lower threshold of excitation and the supramaximal stimulus voltage. The responses may be superimposed and amplified to facilitate taking measurements using variable voltage and timescales in much the same way as a real oscilloscope. Hard copy printouts may also be obtained.

The use of actual experimental data allows realistic-looking, simulated responses to be generated and the biological variation seen in the real experiment, from which the data was generated, is reproduced.

Microcomputer simulations of animal preparations

The approach of animal preparation simulation programs is somewhat different from that of the simulated animal experiment programs. The basis of such programs is to present a simulated animal preparation and to allow the student to design investigative experiments (e.g. the action of a range of drugs/drug combinations or different stimulus parameters). Students are not constrained by teacher-designed experiments and are restricted only by, for example, the range of drugs made available to them by the author. The emphasis here is on students learning by trial and error: they must decide on drug doses and are free to combine drugs however they choose, even if it is inappropriate or would, if performed in the real situation, kill the animal.

Simulated tissue responses are typically generated from a mathematical model which takes into account factors such as drug potency, half-life of response and allows a much greater level of end-user interaction and freedom to design experiments. Since the model may only predict the amplitude and time course of the response, the appearance of the simulated response is often unlike the actual tissue response. While such simulated responses might be aesthetically less pleasing, they do give sufficient information for the student to draw meaningful conclusions from the results. It could also be argued that simulated responses of this sort do reinforce to students the fact that they are using a simulation program.

An example of this type of program is 'Guinea Pig Ileum' (Dewhurst & Meehan, 1989) which simulates the isolated, transmurally stimulated ileum preparation of the Guinea Pig and may be used to investigate the pharmacology of the enteric nervous system. The simulation is based around a much-used pharmacological preparation, where the student is presented with a short length of ileum set up in an organ bath such that contractions of the intestine, either in response to the addition of selected pharmacological agents to the bath or to transmural electrical stimulation, may be recorded. The program is menu-driven and the menu bar places at the disposal of the student a number of items or tools: a stimulator with controls to set the frequency, duration and intensity (voltage) of electrical stimulation; a selection of agents (acetylcholine, clonidine, morphine, naloxone, phentolamine and atropine), each of which may be administered, either as agent-alone or combined with an antagonist, and at a range of concentrations; and a 'magic wash' facility which will instantly remove all trace of any drug added to the bath. Students may design an experiment

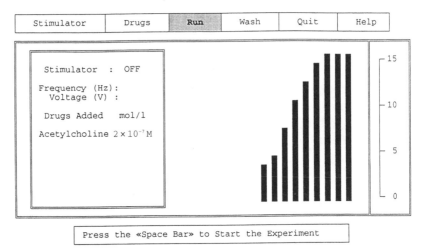

| Stimulator | Drugs | **Run** | Wash | Quit | Help |

Press the «Space Bar» to Start the Experiment

Fig. 14.3 Screen display from 'Guinea Pig Ileum' showing simulated responses of the isolated ileum to different concentrations of acetylcholine administered to the gut-bath. The user may select stimulus parameters (in this instance the stimulator is OFF), a drug from a list of six (in this case acetylcholine) and the drug concentration (in this case 2×10^{-7} M). Results are scrolled from right to left, in real time, and, although it is not clear from the display, the simulated contractions show a rapid rise and a slower rate of fall in tension appropriate to the concentration of acetylcholine administered. Thus here the last of the simulated contractions on the display is the response to the administration of acetylcholine at the concentration shown in the text window. If the microcomputer is connected to a printer a hard copy of the results may also be produced.

and view, on a scrolling display on the monitor, the results (simulated contractions of the guinea-pig ileum) in real time (Fig. 14.3). The experiment may be paused at any time and the experimental variables changed. If the microcomputer is connected to a printer, the results may be printed in real time providing a trace for students to take away and mull over at their leisure.

Students are given little help by the program and additional background or theoretical information is not included. Whilst such freedom of experimental design is desirable, it can be counter-productive in that frequently students are put off if their initial attempts are unsuccessful. Thus, if a student repeatedly selects unrealistic drug doses or inappropriate drug combinations, the 'animal' may not respond (drug dose could be too low) or be killed (drug dose may be too high). In practice, these programs are more successful if students are directed by a teacher-designed practical schedule. Interactive software programs of animal preparations also include: 'Frog Skin', 'Intestinal Absorption', 'Exercise Physiology', 'Squid Axon', 'Ileum', and 'Cardiolab'.

Evaluation of computer-based alternatives

It is important when considering any alternative approach in education to critically compare the new method with the existing approach. In this case, this means evaluating the effectiveness of a computer simulation program in achieving the learning objectives of the practical class on which the simulation is based. In the bio-medical sciences, there has been little evaluation of the problems involved in computer-based learning. The problem is exacerbated by the fact that the learning objectives of practical classes are often poorly defined and rarely made explicit to students (Modell, 1989). A number of the major learning objectives in teaching laboratory practicals in the bio-medical sciences are:

(a) to consolidate existing, and introduce new knowledge;
(b) to develop and practice laboratory skills (generic and specific);
(c) to develop and practice data handling skills (data collection, analysis, presentation and interpretation);
(d) to develop skills in designing experiments;
(e) to practise scientific writing; and
(f) to develop group skills.

Clearly, the second of these objectives cannot be adequately achieved using a computer simulation. However, for many students, probably the majority, it is questionable whether this is of primary importance. Computer simulations of the type described allow students to gather data (which may then be used in the same way as data collected from an experiment) and, in many instances, to develop skills in designing experiments. Indeed, in some universities computer simulations are used to better prepare students who will later perform an animal experiment: they may try out drug doses/drug combinations or ascertain appropriate stimulus parameters. Group skills may also be developed if students are encouraged to work together in solving problems. In one study the student learning experience using computer simulations was found to be enhanced by working in dyads rather than alone (Dewhurst & Meehan, 1993).

There is also evidence that the knowledge gain of students using computer-assisted learning is similar to those taught by conventional methods (Feldman, Schoenwald & Kane, 1989; Hollingsworth & Foster, 1991). Retrospective studies have been carried out in which the performance (measured in terms of the quality of their practical reports) of similar groups of students, some of which used animal experimentation and others used a computer simulation as an alternative, were compared (Clarke, 1987; Dewhurst *et al.*, 1988*a*). Both of these studies demonstrated no difference in student performance. Another study at Wolverhampton University demonstrated that students using an interactive computer-assisted learning program to learn about the physiology of exercise preferred this form of learning to lectures (Coleman *et al.*, 1994).

A more comprehensive study has recently been undertaken (Dewhurst *et al.*, 1994) in which second-year physiology degree students, from the University of Sheffield, were divided into two groups and given the task of investigating nutrient transport in the small intestine as part of a module on epithelia. One group (Control) followed a conventional, closely supervised, laboratory-based approach performing investigative experiments on the isolated, everted intestinal sac of rat, while the second group (Test) worked independently using a computer simulation of this preparation (Dewhurst *et al.*, 1992) in conjunction with a student-centred workbook and problem-solving exercises. The study assessed student knowledge (before and after), student attitudes to computer simulation programs/ practical classes, and compared the resources (staff costs and consumables) required by each approach. The results suggest that students are not disadvantaged by using the computer simulation package. All students were given a written test prior to starting the unit to assess a baseline level of knowledge. This comprised 50 (largely short-answer) questions and included questions requiring descriptive answers, problems to solve, and calculations to perform. After completing the unit they were re-tested and the change in knowledge calculated. The results showed no statistical significance of difference (unpaired t-test; $P = > 0.05$) in knowledge gain (mean gain in knowledge $\pm$S.D.: Control $50.6 \pm 13.0\%$; Test $53.8 \pm 8.5\%$).

A questionnaire survey of student attitudes showed that both the Test group (median score $+11.5$) and the Control group (median score $+1.0$) had a positive attitude to using computer simulations (all students had had previous experience of using similar computer simulation programs) before they undertook the session. The post-session questionnaire analysis showed a statistically significant increase (Mann-Whitney U test: $P = < 0.05$) in positive attitude in the Test group (median $+18.0$) compared to the Control group (median $+1.5$). All students in the Test group showed an increase in positive attitude after using the program indicating that they found it a useful learning aid. The attitude of the Control group, none of whom used the program (although they were given the option to do so in their own time), remained unchanged.

The resource analysis took into account: consumables (e.g. reagents, animals, disposable apparatus), and staff resources (academic staff time costed at £40 per hour, technician time costed at £10 per hour, demonstrator time costed at £10 per hour, and glasswashing time costed at £5 per hour) required for the two approaches. Capital cost of equipment for both groups was ignored. This analysis demonstrated quite clearly that using computer simulations is less expensive: the laboratory-based approach was approximately ten times more expensive than that of the computer simulation (approximately £170 less expensive per student). A similar study with medical students, carried out in the Department of Pharmacology, Charing Cross & Westminster Hospital, demonstrated similar findings (Leathard *et al.*, 1994).

Conclusions

Views on the use of animals in education are sharply polarized, and it is difficult to see a significant change in the current situation at least in the short term. However, a number of factors suggest that vivisection in the classroom will decrease. First, animal rights issues currently have a high media profile raising student awareness and promoting a change in attitude to vivisection. Secondly, the drive to reduce unit costs in higher education has made managers question the educational benefits of practical classes which are extremely resource intensive. The importance of economic constraints should not be underestimated. Thirdly, a range of alternatives now exist which previously were not available to teachers. These include a move towards the use of human subjects in investigative laboratory practicals instead of animals, the use of plant rather than animal tissue in biochemistry practicals, and the use of video, models and interactive video as alternatives to animal dissection. However, perhaps the greatest impact has been the use of computer-based alternatives in undergraduate physiology and pharmacology courses. The increased availability and processing power of microcomputers together with the introduction of more sophisticated software authoring packages has made it possible to harness technology to produce interactive programs which simulate animal experiments or animal preparations and which students seem to readily accept. A number of studies have demonstrated such alternatives to be equally valid educationally and, in many instances, more cost effective. For many students, computer simulations of the type described may well achieve most of the learning objectives of laboratory practical classes. Clearly, laboratory skills especially those specific to a particular animal preparation cannot be simulated. However, for many students, probably a majority, it is questionable whether such skills are important. It might be argued that such skills are only really important for those students intending to pursue a career in research. The pharmaceutical companies, who are major employers of pharmacology graduates, have expressed concern about the lack of hands-on experience of animal work in some university courses. It is the responsibility of the teacher to better define the learning objectives for different groups of students and not simply adopt the policy that animal experiments are essential for all. Clearly, the acquisition and practice of generic laboratory skills is important for all students of biology though these should be addressed in other laboratory exercises not involving animals.

The use of computer simulation programs will undoubtedly be different for different groups of students. One possible role may be seen simply as that of better preparing students who will go on to perform the experiment *in vivo*, where the simulation is used, for example, to try out drug doses and/or drug combinations or test different stimulus protocols. Alternatively, they could be used as a 'fallback' situation for those students who fail to obtain results. A third, and increasingly attractive option is that they may provide a realistic, animal-free, non-invasive and less expensive but educationally equivalent alternative to the real experiment. The importance of evaluating any new approach cannot be overstressed. If teachers are to be persuaded to alter their teaching methods, they must be convinced of the educational advantages and the cost effectiveness.

Acknowledgements

I am greatly indebted to the hard work and expertise of a team of co-workers including Tony Meehan, Megan Quentin-Baxter, Clint Howells, Alan Williams, Guy Brown and Richard Ullyott without whom none of these programs would have been developed. I would also like to thank The Lord Dowding Fund (NAVS, UK) for generous financial support.

References

Barker, S. P. (1988). Comparison of effectiveness of interactive videodisc versus lecture-demonstration instruction. *Physical Therapy*, **68**, 699–703.

Basarab-Horwath, I. , Dewhurst, D. G. , Meehan, A. S. & Odusanya, S. (1986). Analysis of evoked electromyographic activity using the BBC microcomputer. *Journal of Physiology (London)*, **378**, 3P.

Bender, D. A. (1989). Combining a computer simulation with a laboratory class – the best of both worlds? *Computers in Education*, **13**, (3), 235–43.

Brown, G. J. , Collins, G. G. S. , Dewhurst, D. G. & Hughes, I. E. (1988). Computer simulations in teaching neuromuscular pharmacology – time for a change from traditional methods? *Alternatives to Laboratory Animals*, **16**, 163–74.

Brown, E. R. , Dewhurst, D. G. & Williams, A. D. (1994). An interactive computer simulation of the squid giant axon preparation for teaching undergraduate students. *British Journal of Pharmacology Proceedings Supplement*, **112**, 437P.

Clarke, K. A. (1987). The use of microcomputer simulations in undergraduate neurophysiology experiments. *Alternatives to Laboratory Animals*, **14**, 134–40.

Clarke, K. A. (1988). Microcomputer simulations of mechanical properties of skeletal muscle for undergraduate classes. *Alternatives to Laboratory Animals*, **15**, 183–7.

Coleman, I. , Dewhurst, D. G. , Meehan, A. S. & Williams, A. D. (1994). A computer simulation for learning about the physiological response to exercise. *American Journal of Physiology*, **266** (Advances in Physiology Education, Vol. II), S2–S9.

Croall, J. (1994). The Cutting Edge: Schools: Animals in Science. (1994). *Guardian Education* March 15th.

Dewhurst, D. G. & Howells, C. (1990). A computer simulation of the cat nictitating membrane preparation (*in vivo*) for teaching undergraduate pharmacology students. *Alternatives to Laboratory Animals*, **17**, 291–300.

Dewhurst, D. G. & Meehan, A. S. (1989). Computer simulation of the effects of drugs on neurotransmitter release in the enteric nervous system. *British Journal of Pharmacology Proceedings Supplement*, **97**, 597P.

Dewhurst, D. G. & Meehan, A. S. (1993). Evaluation of the use of computer simulations of experiments in teaching undergraduate students. *British Journal of Pharmacology Proceedings Supplement*, **108**, 238P.

Dewhurst, D. G. & Ullyott, R. T. (1991). Computer simulated demonstration of the actions of drugs on the isolated perfused mammalian heart (Langendorff preparation). *Alternatives to Laboratory Animals*, **19**, 316–22.

Dewhurst, D. G. & Williams, A. D. (1993). Frog skin: a computer simulation of experiments performed on frog skin *in vitro* to investigate the epithelial transport of ions. *Alternatives to Laboratory Animals*, **21**, 350–8.

Dewhurst, D. G. , Brown, G. J. & Meehan, A. S. (1988*a*). Microcomputer simulations of laboratory experiments in physiology. *Alternatives to Laboratory Animals*, **15**, 280–9.

Dewhurst, D. G. , Brown, G. J. & Meehan, A. S. (1988*b*). Computer simulations – an alternative to the use of animals in teaching. *Journal of Biological Education*, **22**(1), 19–22.

Dewhurst, D. G. , Hardcastle, J. , Hardcastle, P. & Williams A. (1992). An interactive computer simulation of experiments to teach the principles of nutrient transport in the small intestine. *Alternatives to Laboratory Animals*, **20**, 529–35.

Dewhurst, D. G. , Hardcastle, P. T. , Kohn, P. , Stuart, E. & Williams, A. D.

(1993). A computer-assisted learning package to teach intestinal nutrient transport to undergraduate students – an evaluation. *Journal of Physiology (London)*, **467**, 224P.

Dewhurst, D. G. , Hardcastle, J. , Hardcastle, P. T. & Stuart, E. (1994). Comparison of a computer simulation program with a traditional laboratory practical class for teaching the principles of intestinal absorption. *American Journal of Physiology* (*Advances in Physiology Education*), (in press).

Downie, R. & Meadows, J. (1995). Experience with a dissection opt-out scheme in university level biology. *Journal of Biological Education*.

Feldman, R. D. , Schoenwald, R. & Kane, J. (1989). Development of a computer-based instructional system in pharmacokinetics: efficacy in clinical pharmacology teaching for senior medical students. *Journal of Clinical Pharmacology*, **29**, 158–61.

Guy, J. F. , Frisby, A. J. (1992). Using interactive videodiscs to teach gross anatomy to undergraduates at The Ohio State University. *Acadamic Medicine*, **67**, 132–3.

Hollingsworth, M. & Foster, R. W. (1991). Computer-assisted learning (CAL) programs in drug disposition – replacement of lectures. *British Journal of Pharmacology Proceedings Supplement*, **102**, 216P.

Hughes, I. E. (1984). The use of computers to simulate animal preparations in the teaching of practical pharmacology. *Alternatives to Laboratory Animals*, **11**, 204–13.

Hughes, I. E. (1987). Computer simulation of cardiovascular responses from *in vivo* preparations. *British Journal of Pharmacology Proceedings Supplement*, **90**, 290P.

Langley, G. R. (1991). Animals in science education – ethics and alternatives. *Journal of Biological Education*, **25**, 274–9.

Leathard, H. L. , Cover, P. O. , Dewhurst, D. G. , Kumari, M. & Rantle, C. (1994). Evaluation of a computer assisted learning program on intestinal motility. *British Journal of Pharmacology Proceedings Supplement*, **112**, 179P.

Lock R. (1993). Animals and the teaching of biology/science in secondary schools. *Journal of Biological Education*, **27**, 112–4.

Lock, R. & Millett, K. (1992). Using animals in education and research – student experience, knowledge and implications for teaching in the National Science Curriculum. *School Science Review*, **74**, 276–80.

MacDonald, E. & Saano, V. (1987). Video teaching – glamour without effect. *Trends in Pharmacological Sciences*, **8**,(3), 77–8.

Modell, H. L. (1989). Can technology replace live preparations in student laboratories? *American Journal of Physiology*, **256** (Advances in Physiology Education), S18–S20.

Molloy, J. E. , Newport, A. P. & White, D. C. S. (1986). An inexpensive method for capturing, storing and analysing data from class neuro-physiology experiments using a BBC microcomputer. *Journal of Physiology (London)*, **376**, 11P.

Nab, J. (1990). Reduction of animal experiments in education in The Netherlands. *Alternative to Laboratory Animals*, **18**, 57–63.

Operation Frog (1984). Scholastics Software, Scholastic Inc. , 730 Broadway, New York, NY 10003, USA.

Orlans, F. B. (1988). Should students harm or destroy animal life? *American Biology Teacher*, **50**, 6–12.

Quentin-Baxter, M. & Dewhurst D. G. (1992). An interactive computer-based alternative to performing a rat dissection in the classroom. *Journal of Biological Education*, **26**,(1), 27–33.

Quentin-Baxter, M. & Dewhurst, D. G. (1995). An interactive video laser disk to teach the functional anatomy of rat. *Journal of Biological Education*.

Smith J. (1992). Dissecting values in the classroom. *New Scientist*, **134**, 31–5.

Statistics of Scientific Procedures on Living Animals in Great Britain for 1987; 1988; 1989; 1990; 1991; 1992: UK: HMSO.

Webster, E. & Myers, J. M. (1986). A simple ECG monitor using fibre optic isolation. *Journal of Biological Education*, **20**,(2), 99–102.

15

Animals in scientific education and a reverence for life

HENK VERHOOG

A contradiction in the goals of biology education

In many countries one of the goals of biology education is to promote respect for all forms of life. Bowd (1993, p. 84) refers to the National Science Teachers Association (1980) in North America, stating that the central affective goal in the secondary science education curriculum is that learning should engender a 'reverence for life'. He brings up the question of whether killing and dissecting animals for educational purposes is not contradictory to this aim. This chapter will present the inbuilt tension between this goal of biological education and the emphasis within modern biology on the objectification of nature, including the experimental reductionistic methods of studying living nature. Commenting on Bowd, Lock (1994, p. 76) writes that 'It is much more common now for animals to be dissected and used for an investigation directed towards an understanding of the process of scientific enquiry', rather than as an exploration in anatomy. Here lies the heart of the problem: to understand the role of dissection and animal experimentation in biology, we have to understand the process of scientific enquiry.

When teaching students about the methodological, social and ethical aspects of biology, it is important to develop a conceptual framework which is effective in sensitizing students to the philosophical aspects of biology. In the framework used by the author various biological disciplines are looked at as operating between two imaginary poles, the poles of 'given nature' and of 'constructed nature'. 'Given nature' refers to nature which exists autonomously, independent of human influences ('wild' nature). The other pole of 'constructed nature' can best be visualized when we go into a modern laboratory where molecular biologists work. What we experience directly as 'nature' in the world of everyday life has become almost invisible here.

All science once started with nature as we experience it in our world of everyday life. Although scientific knowledge has already influenced our concept of nature in everyday life to a very large extent, it can still be maintained that nature as experienced by humans has more dimensions than the ones investigated by natural science. Two more obvious dimensions are the aesthetic and the spiritual, religious dimension, which play such an important role in environmental ethics and environmental philosophy.

The manner in which nature is conceptualized and studied in the experimental natural sciences is just one perspective in examining nature, a way which has become very dominant in our society. To understand the role of animal dissection and animal experimentation in biology we must find out what is characteristic of this dominant scientific approach to nature. In the context of the framework mentioned before the question becomes: what happens in the scientific process of transforming 'given nature' into constructed nature?

In this chapter, two topics will be discussed. The first topic will be the objectification of nature and the role of experimentation in scientific enquiry. The second topic will present a discussion of the moral and educational implications of this analysis.

The objectification of nature

Natural science wants to have knowledge about nature which is 'objective'. In the context of natural science, what is considered to be objective is opposed to that which is considered to be subjective. The meaning of these two terms is dependent upon the way we look at the relation between subject and object. For knowledge to be at all possible, we have to make a distinction between a knowing subject and the object to be known. The object to be known can be anything which appears in our perception; including inner experiences such as dreams, feelings, thoughts, etc. In the standard conception of natural science certain limitations are introduced as to the objects to be studied. Only those objects are legitimate objects of scientific study which are intersubjectively visible for human eyes, or can be made visible, even if in a very indirect way. Ultimately this goes back to Descartes' ontological dichotomy between the immaterial world of the subject ('*res cogitans*') and the material world of nature as object ('*res extensa*').

The materialistic concept of nature, given with the Cartesian dichotomy, creates an almost unbridgeable gap between man as subject, and nature as material object. In the world of nature studied by science there is no place for a human or animal subject, or the qualities characterizing these subjects. The characteristics of an experiencing subject are not admitted in the description and explanation of the natural world. Another way to illustrate this is the distinction between 'secondary' and 'primary' qualities. In line with the Cartesian dichotomy primary qualities are considered to be the 'objectively' real properties of the outside world, existing independent of an experiencing subject. Primary qualities are the measurable and quantifiable properties of Descartes' *res extensa*, the real properties of physical nature. Secondary qualities, on the other hand, are the properties of nature as they are experienced subjectively, such as colours. In this view, also the experiences related to other dimensions of nature, say the aesthetic or spiritual, are seen as 'subjective'. These experiences say something about ourselves as subjects, not about the objective world.

The characteristic way of objectification of natural science can explain why the question whether animals are conscious is such a hot issue. Strictly speaking, the same applies to human consciousness. The dualistic view allows introspection in our own conscious experience, but makes it impossible to have direct access to the consciousness of others, whether human or animal. It is here that the distinction between nature as experienced in our world of everyday life, and physical nature (nature objectified as described) can be experienced very vividly. Wieder (1980) describes how researchers studying animal behaviour in the 'behaviouristic' tradition deal with chimpanzees outside of the experimental context. The chimpanzees are then dealt with as experiencing subjects, as embodied consciousness, and not as material objects (mere bodies). In this community, subjectivity is apprehended directly in face-to-face encounters. Once these animals become objects of research, however, all subjective references are truncated. A new order of events is created, the order of pure objectivity that stands over and against the order of everyday life.

Rollin (1990, Ch. 1) describes in detail how, in connection with the study of animal consciousness, science has become increasingly remote from common sense and ordinary experience. Also that science rests on its own philosophical and ideological presuppositions, which are rarely examined. These presuppositions determine what counts as real, as fact, as legitimate data and explanations. According to these presuppositions, data about conscious experience or animal mentation are not considered as legitimate data; the mental phenomena have to be reduced to neurophysiological or chemical data. The tension between human experience in the world of everyday life and laboratory data is present everywhere in Rollin's book.

The direct and often anecdotal evidence of animal consciousness and animal pain in our everyday life is denied in the very artificial setting of the laboratory, where the first aim of research is control. Also, the moral relevance of experiential data about animal consciousness is disappearing in the laboratory setting; the objectification of science leads to the separation of science and ethics.

The animal as laboratory equipment

Rollin (1990) describes how the presuppositions mentioned before were introduced at the end of the last century. The outcome was that the study of animal minds (i.e. subjective experiences) was considered to be unscientific and scientifically impossible. It was also during this period that the experimental method was introduced into the biomedical sciences. Bernard was one of the first scientists who explicitly emphasized the importance of this method in order to achieve what he considered to be the goal of science: the production of data which can be repeated by others. The controlled conditions in the laboratory led to greater certainty, quantification and thus, predictability. Within the confines of the laboratory the conditions can be far better controlled than in the natural world.

In historical perspective the role of the animal as the 'object' of biological research changes significantly during the 19th century. Gersch and Gersch (1977) describe how, before the 19th century, the emphasis of the biologist was on the animal as an indivisible whole. As an object of description and classification, the animal was studied for its own sake. After the introduction of the analytical experimental method, the animal as an organic whole was destroyed and animals were more often used as means to answer theoretical and practical questions. The animal became an instrument, a model, research material, and it was the task of laboratory animal science to perfect these models. Gersch and Gersch (1977) write about the increasing standardization of animal models to increase the reliability and repeatability of the results. Their paper vividly emphasizes the contrast between the direct experience of spontaneity and variability as characteristics of animal behaviour and the ideal of creating laboratory animals whose behaviours can be controlled and made predictable.

With this change in the object of biological research, there was also a change in the definition of objectivity, as will be illustrated by the work of Hearne (1987). As was shown before in connection with the study of animal consciousness, there is a discrepancy between our experience in the world of everyday life and scientific objectification of animals. Hearne (1987) is a professional trainer of dogs and horses. Entering academia with an interest in philosophy, she was surprised to find that professors specifically denigrated students' language describing animals in subjective terms, that is, anthropomorphically. In Hearne's opinion the anthropomorphic language of everyday life, the language which is also used by trainers, is true to the nature of the animals as we experience them, and in that sense perfectly objective. She refers here to the definition of objectivity as 'being true to the nature of the object studied'.

The changes at the end of the last century, as described by Rollin (1990) and Gersch and Gersch (1977), were not based on new empirical evidence or experiments but involved the epistemological choice to apply the Cartesian criteria of scientific objectivity to the study of living nature. This choice is then reinforced by the implementation of the experimental method. In most cases the application of this method involves an analytical and reductionistic approach to animals, in which the criterium of objectivity is transformed into 'repeatability', preferably under controlled laboratory conditions.

It is only recently that philosophers and sociologists of science show interest in the role of the experiment in creating the 'facts' of natural science. The work by Lynch (1988) and Arluke (1988, 1992) illustrates this new development. Lynch distinguishes between the 'naturalistic' animal of the common sense perspective and the 'analytic' animal of the laboratory scientist. The naturalistic animal is the subject of anthropomorphic identifications. In the process of research this animal is transformed into the analytic animal, into data. In the scientific system of knowledge the analytic animal is seen as the real animal. According to Lynch (1988, p. 270) the laboratory procedures as such 'assure the removal of the characteristics that make up the naturalistic animal'. How this is done and the ambiguities involved in this process are also described by Arluke (1988, 1992). He (1988, p. 100) writes about 'counter-anthropomorphism' when inanimate qualities are attributed to living things. Analytic animals are de-individualized and treated as anonymous beings. Social norms in the laboratory prevent scientists and animal technicians from treating laboratory animals as pets; they are instead treated as models, as supplies in grant proposals, etc. Arluke (1992, p. 34) believes that this objectification or detachment is necessary for self-protection. Objectification breaks the interconnectedness between subject and object, thus moral constraints are nullified. Arluke (1992) suggests that this process of objectification rarely succeeds completely. Many animal experimenters have emotional difficulties with invasive animal experiments. However, in the laboratory setting there rarely is the possibility of openly discussing these problems. In the laboratory, in general, feelings about animal use remain private and extraneous to the 'real work' of the laboratory.

Moral and educational implications

The epistemological presuppositions and experimental methodology of natural science imply a denial of animals as subjective beings. When animals are seen as anonymous objects in the process of scientific experimentation, there is no necessity for examining the ethics of their use. Arluke (1992, p. 35) found that people in the laboratory 'did not have elaborate moral justifications for their use of animals. Instead, many of them appeared ethically inarticulate'. Animal experimentation thus corresponds with a very anthropocentric attitude towards animals, as if the use of animals for scientific research needs no further legitimization. This is further reinforced by the idea that science has to be value-free and objective, that science has to describe nature 'as it really is', uncontaminated by human subjective feelings. That is the reason why the feelings of the animal experimenters have to remain 'private'. It also explains why students who object to the use and killing of animals for educational purposes are accused by their teachers of being too emotional, sentimental and irrational. Science is considered to be objective and rational, ethics subjective and irrational. Animal experimentation is believed to be inextricably connected with studying biology, and conscientious objectors are usually advised not to study biology. Given the philosophical underpinnings of scientific theory and the subsequent focus of scientific education, is it really possible to provide students using animals in their education a climate in which the goal of promoting respect for all life can be reached? It is difficult to see how this might be the case when the very science that claims to be 'objective' and value-free, ignores an animal's subjective experiences as well as those of the students and researchers.

The current status of animals in experimental biology is the result of a particular historical and social process. Animals used to be treated in the way biologists have done so since the last century, according to accepted social and methodological practices: animals are considered to be research and educational tools; their subjective well-being is not considered nor are the subjective/empathetic experiences of the researcher/student taken into account during the process of scientific investigation. Meanwhile this view of science has been criticized very much, especially since the Positivist tradition in scientific philosophy and sociology, as described by Rollin (1990), has been replaced by more Relativist and Post-Modern views. In these latter views it is denied that criteria exist by which it can be decided what is true and value-free knowledge, totally independent of any social and historical context (Knorr-Cetina & Mulkay, 1983). In these more relativistic views, natural science is seen as a particular social manifestation of the relation between man and nature, in which human control over natural processes is the main goal.

It is because of the experimental approach, in which nature is studied under controlled laboratory conditions, that the methodology of natural science and of modern technology are intrinsically related, and therefore it is difficult to see how science can be value-free. One could even say that the better scientists succeed in controlling nature under laboratory conditions (the more 'objective' the results are in terms of repeatability and predictability), the more useful the results of science may be for technological control of the world (Verhoog, 1980). The developments in modern biotechnology show this very clearly.

Another reason not to accept the value-freedom of science would be the acceptance that animals have a moral status or intrinsic value (Verhoog, 1992). The scientific method is only ethically neutral as long as one denies that the organisms studied by the natural scientist have a value of their own, a value which is independent of their instrumental value for human purposes. Once the intrinsic value of animals is accepted, any use of animals by humans has to be legitimized; the scientist must give good ethical reasons for experimenting with animals.

In the Netherlands, the idea of the intrinsic value of animals has been introduced in the laws dealing with animal well-being. This has had a number of implications. Special ethical committees have been designed to ensure that the ethical reasons put forward by the scientists are valid. In this way, scientists who never before considered the ethical considerations of animal use are now forced to do so. It is no longer the student, but the teacher in science who has to explain to students what the ethical reasons are why s/he thinks that animals may be killed or used in experiments. The teacher is required to respect students who have a different opinion about the invasive use of animals for the purpose of education (Verhoog, 1993). In discussions with students, the question of animal subjectivity, and the world of everyday life experience of students is considered to be important. The teacher/professor must discuss the costs and benefits, emotional and practical, of animal use in education. There should also be a presentation to the student about the extent of animal sacrifice involved in experimentation and the possibility of experimental alternatives. When animals have intrinsic value, an experiment should not be done if there is an alternative way of reaching the goal, without using animals.

Lynch (1988) and Arluke (1988, 1992) have described the laboratory process of transforming the 'naturalistic animal' into an 'analytic animal', an example of the tension existing within biology between 'given nature' and 'constructed nature'. There is evidence that the relationship between science and our moral beliefs about animals is changing the further away we get from the direct experience of nature. If this is true, then there is reason to believe that fostering an attitude of respect and reverence for life comes from those biologists who remain fairly close to 'given nature' and study the 'naturalistic animal' (e.g. systematic biologists, field ecologists, ethologists, morphologists). In scientific education and research there is a strong tendency now to emphasize those fields of biology which are closer to 'constructed nature'. Many introductory biology textbooks begin with cell-biology, molecular biology and genetics. Organismic biology and especially ethology and ecology are placed at the end of the textbook. A heavy emphasis on reductionism pervades biological research and education. This is accomplished through accentuating the belief that phenomena at higher levels of biological organization have to be explained by reducing them to lower levels of organization. If the goal is to encourage students to have a respect for life, then the reductionist approach must be kept in balance with that of the holistic approach (i.e. examining whole animals and systems). The value of the data reached by reductionist analysis depends on the data placed within a larger whole, a higher level of organization, ultimately the whole animal, its behaviour and interaction within the greater ecosystem of that particular species. When students have internalized respect for life, the next step may be reductionist analysis, but it must be made clear from the very beginning that this often involves a violation of the integrity and well-being of animals, and that one must have very good reasons to sacrifice animals for the purpose of scientific research or scientific education.

An excellent example of research maintained in balance between the whole plant and its genetics, is McClintock (Keller, 1983). For McClintock, the goal of science was not prediction (based on controlled laboratory studies) but understanding based on our connection with the world. For her the secret of science was looking and listening, listening to what nature wants to tell us. She talked about chromosomes as if she knew what it is to be a chromosome, as if the chromosome became a partner in a dialogue. To achieve this, one has to forget oneself, to let the book of nature speak within oneself. The work of McClintock shows that a reductionist approach does not necessarily mean that one has to ignore the plant or animal within the 'whole' context.

It is no wonder that the more phenomenological way of doing science of McClintock goes together with a great respect for the intrinsic value of nature, of life. Teaching students to respect life means fostering such an attitude in the process of education. The Swiss comparative zoologist Portmann (Grene, 1968) also worked in a holistic, phenomenological tradition. He was especially interested in the ontology and morphology of animals, both invertebrates and vertebrates. Portmann was very much aware of the growing gulf between the biotechnical approach in biology and the comparative holistic approach. In the latter approach one looks to the 'authentic' phenomena, directly visible to the human senses, and compares these phenomena in order to find out the meaning within a larger whole. The other, reductionistic approach, leads to causal analysis, to discover the mechanisms behind the scene, as it were, the mechanisms which lead to control over nature. Portmann (Grene, 1968) made a distinction between a 'theoretical' and an 'aesthetical' function of the human mind. In the theoretical function, the mind tries to transform qualitative phenomena to quantitative data by means of rational thinking. In the aesthetic approach, the scientist does not try to explain what is directly given in our perception by means of mechanisms which are postulated inside an organism. Such as aesthetic and more intuitive perception of organisms can, according to Portmann teach us something about what is 'essential'. The concept of intrinsic value is directly related to what is essential for the life of the organism involved. It sensitizes us to what is real about animals, independent of the use we may have of them. It is necessary to question the reasons for experimenting on animals to determine if the answers are satisfactory within the boundaries of a moral stance and a basic respect for life.

In the reductionist-experimentalist approach, based on a Cartesian dualistic viewpoint, the knowledge of the primary qualities is considered to be knowledge of the real world, and the secondary qualities are subjective. Portmann takes the opposite view. What we experience in our world of everyday life comes first, is the primary world, and what the scientist finds in the laboratory is secondary, is derived. Portmann was concerned about the fact that in biology the emphasis shifted more and more to the causal-analytic approach. He was convinced that further alienation from nature, which is the result of this latter approach, can only be prevented when people have a direct experience of the richness, of the diversity of biological form, especially in scientific education. The reverse occurs in most programmes of scientific education, where the emphasis is on molecular biology and genetics, probably because these subjects are now considered to be the 'cutting edge' of science. Maienschein (1994) criticizes this development, stating that a particular conception of what is to be considered 'good science' underlies it: science must be analytical, must control as many variables as possible, and it must produce solid and unquestionable positive results. Maienschein's (1994, p. 22) conclusion is:

The present preoccupation with the cutting edge is unfortunate for biology. The dominance which some areas of biology have been allowed to assume because of the acceptance of a concept of a cutting edge is pernicious.

If we want to maintain and promote a respect for life as an important goal in the teaching and practice of biology and research, then reductionist principles as a guiding conceptual framework within science should be presented in parallel with that of a holistic framework. In this way, students should be enlightened regarding the variability in scientific methods available to them. The knowledge to do this is available (Von Gleich, 1989); there is no reason to wait any longer.

References

Arluke, A. B. (1988). Sacrificial symbolism in animal experimentation: object or pet? *Anthozoös*, **II/2**, 89–117.

Arluke, A. B. (1992). Trapped in a guilt cage. *New Scientist*, **4 April**, 33–5.

Bowd, A. B. (1993). Dissection as an instructional technique in secondary science: choice and alternatives. *Society and Animals*, **1/1**, 83–9.

Gersch, M. & Gersch, D. (1977). Das Objekt in der biologischen Forschung. *Biologische Rundschau*, **15**, 145–60.

Gleich, A. von (1989). *Der wissenschaftliche Umgang mit der Natur. Über die Vielfalt harter und sanfter Naturwissenschaften*. Frankfurt/New York: Campus Verlag.

Grene, M. (1968). Approaches to a Philosophical Biology. New York/London: Basic Books.

Hearne, V. (1987). *Adam's Task. Calling Animals by Name*. London: Heinemann.

Keller, E. F. (1983). *A Feeling for the Organism*. New York/Oxford: Freeman and Company.

Knorr-Cetina, K. D. & Mulkay, M. (1983). *Science Observed. Perspectives in the Social Study of Science*. London/Beverly Hills: Sage Publications.

Lock, R. (1994). Dissection as an instructional technique in secondary science: comment on Bowd. *Society and Animals*, **2/1**, 67–73.

Lynch, M. E. (1988). Sacrifice and the transformation of the animal body into a scientific object: laboratory culture and ritual practice in the neurosciences. *Social Studies of Sciences*, **18**, 265–89.

Maienschein, J. (1994). Cutting edges both ways. *Biology and Philosophy*, **9**, 1–24.

Rollin, B. E. (1990). *The Unheeded Cry. Animal Consciousness, Animal Pain and Science.* Oxford/New York: Oxford University Press.

Verhoog, H. (1980). *Science and the Social Responsibility of the Scientist.* Meppel: Krips Repro.

Verhoog, H. (1992). The concept of intrinsic value and transgenic animals. *Journal of Agricultural and Environmental Ethics,* **5/2,** 147–60.

Verhoog, H. (1993). Animals in education and the structure of science. *Global Bioethics,* **6/3,** 177–85.

Wieder, D. L. (1980). Behavioristic operationalism and the life-world: chimpanzees and chimpanzee researchers in face-to-face interaction. *Sociological Inquiry,* **50/3-4,** 75–103.

Part V

Epilogue:
the future of wild animals

16

Human sentiment and the future of wildlife

DAVID E. COOPER

How wildlife is to be managed in the future depends, in large part, on why it is thought important for wildlife to have a future.[1] Someone who thinks in terms of the medical or photographic opportunities which wildlife provides is liable to advocate policies different from those of a person who appeals to ecological balance. People to whom the fates of individual animals matter are unlikely to pursue exactly the same policies as those for whom it is the species that really counts.

The management issue, then, turns on broadly moral considerations which are not the preserve of scientists and experts in the way that the implementation of proper policies may be.

Unfortunately, the blunt moral questions, 'why ought there to be wildlife? why would it be wrong to allow its demise?' have a certain intractability. To begin with, 'wildlife' is a vague term. Does it, for example, apply to the deer herds of Nara or Richmond Park? It is also a huge category, embracing several million species, some 50 000 vertebrates included. It is not obvious that the demise of mussels or termites would be wrong for the same reasons as that of dolphins or leopards. Intractability is due most of all, however, to the immaturity of our moral thinking on such matters. There have always been individuals, from the Buddha to Bernard Shaw, concerned about our treatment of wild animals, but it is only with the recent massively visible threat to wildlife that serious efforts have been made to construct a 'wild life ethic' as an extension to that seasoned moral thinking whose compass, hitherto, has been human beings alone.

---

[1] Some people dislike reference to wildlife *management*, and understandably so, since it smacks of the 'dominion' attitude to animals. But the phrase is now common currency and here to stay. Let me make clear that, as I use the phrase, 'management' could as well apply to a 'hands off' policy of letting animals get on with living in wildernesses or oceans as to one of corralling them in 'safari parks' or 'sea-worlds'.

Our 'human ethic', seasoned as it is, still allows, of course, for large disagreements: nevertheless, there is some consensus as to why we feel it important to have principles for the decent treatment of our fellows. It is because we are, in David Hume's words, creatures with a 'limited sympathy' for one another. We need, therefore, orderly outlets for this sympathy, and ways of reducing the antagonisms to which the limitations on that sympathy give rise. It is far from clear what the analogous basis for a 'wildlife ethic' might be. After all, in our century, at least, antagonisms between humans and animals are settled, rather easily, in favour of the former. Power, not principle, is sufficient to resolve such conflicts.

It is with the more tractable questions of why many of us feel it is important that wildlife should be 'morally considerable' and why we would regard it as something dreadful if certain species were denied a future that I engage. The questions are ones about a human sentiment concerning these species. Unless that sentiment is diagnosed, little progress can be made towards answering the direct question of why it is 'really' wrong to aid and abet the extinction of wildlife – just as, I would suggest, progress in devising a 'human ethic' requires clear appreciation of that 'limited sympathy' on which standards of reciprocal behaviour are founded. (I refer, you will have noted, to a sentiment concerning certain species. This is because few people could claim, hand on heart, to view the prospect of a cockroach-less world with the same gloom as they do that of a world without, say, elephants and dolphins. This is not to suggest that cockroaches don't matter at all, only that any understanding of why they might must refer back to a sentiment which does not normally include them in its scope.)

In this paper, I argue that the most familiar reasons given for concerning ourselves with wildlife fail to do justice to the sentiment which inspires such concern, and therefore fail to provide effective platforms from which to speak on behalf of wildlife. In so arguing, I diagnose the nature of this sentiment and consider the management and educational policies that an ethic consonant with that sentiment might favour.

Four kinds of reasons dominate the literature on concern for the future of wildlife. The first appeals to the rights of individual animals: hunting rhinos violates the rights of actual rhinos and those of the rhinos which would have existed if their ancestors had not been exterminated. The second also appeals to rights, but this time to the rights of species to exist and flourish. A species, some say, is a 'vital life-stream' which, in a good world, is allowed to flow.[2] Besides these 'zoocentric' reasons, there are those which directly appeal not to the interests of animals, but to ones much narrower or much broader. The first of these argues for wildlife concern on the basis of its benefits for just one species, human beings. The other rests this concern on the benefits for whole environments, or ecosystems, or for 'Gaia' perhaps.

There are, of course, other arguments heard – those, for instance, which speak of duties to God's own creatures; but the four mentioned above are the most familiar. It is worth briefly discussing their interrelations, but it needs to be stressed that each argument can appear in various forms. Advocates of animal rights, for example, can disagree as to exactly what rights individuals or species possess (to existence? to freedom? to a natural environment?), and as to the grounds, if any, for overriding these rights (hunger? self-defence? *Lebensraum*?). Arguments from human interests may cite anything from unusual meat products to aesthetic delight in the grace and form of some animals. Again, those whose concern is more global may offer various criteria for the well-being of the environment to which wildlife contributes: for instance, ecological balance and maximum diversity.

Considered as final courts of appeal, the four kinds of reasons are incompatible, for even if they converge on a policy, the grounds for the policy are quite different. Whether or not they do converge depends on the particular versions proposed and on assessments of the empirical facts. All four might agree in condemning fox-hunting, for example: it causes individual suffering: it threatens the survival of the foxes; it encourages brutal attitudes, dangerous to other people, among its aficionados; and it disturbs nature's balance. Equally, all four might agree to support, or condone, fox-hunting: it's better than leaving foxes to the mercies of the terrier-men; it encourages the fox population, since farmers profit by sparing the animals for the Hunt; it's very enjoyable and a welcome gesture of defiance against a society dangerously bent on destroying our traditions; and, by inducing farmers to grow hedges and woods, it encourages a variety of wildlife in the vicinity of the Hunt. By permitting the various considerations mentioned, it is easy to produce conflicting verdicts on fox-hunting among spokesmen for the four positions.

On more general principles of wildlife management, the four kinds of reasons will tend to support different approaches. Other things being equal, advocates of individual animals' rights will be less sympathetic to management within zoos and 'safari parks' than those for whom the

---

[2] See, for example, Holmes Rolston III, 1986, Ch. 10.

imperative is species conservation. *Prima facie*, those whose final concern is with human beings only will be less enthusiastic for recovering virgin wildernesses than those who speak the language of 'planetary health'. But I use terms like 'tend', 'other things being equal', and *prima facie*, since the four positions are each too amorphous, and the empirical facts too uncertain for definite implications for practice to be drawn.[3]

None of the positions, I suggest, in any of their versions, explains the judgement that there is tragedy – and a moral debacle[4] – in the continuing erosion of wildlife. None of them properly reflects the sentiment which underlies that judgement. I confine myself to one critical point against each.

(a) People sympathetic to the vocabulary of individual animal rights will, of course, deplore the violation of these rights which the manner of wildlife decimation often involves, for example, the capture of baby chimps and the accompanying murder of their mothers.[5] But, since what offends the sentiment I am after is also the idea of a future stripped of wildlife, it cannot be solely the harm inflicted on actual animals which constitutes the offence. It is, so to speak, the failure of animals to exist in that future which disturbs. Now it is notoriously hard to explain what is wrong with such a future in terms of violating individual rights, for appeal would have to be made to the notion, perhaps an unintelligible one, of the rights of creatures which might have existed, but actually will not. Even if sense can be made of a merely 'possible animal's' right to exist, the notion is surely too subtle to explain the familiar feeling that a world without wildlife would be a dreadful one. Or, to put it like this: the effect of understanding what would be dreadful about it in terms of violation of rights is to regard it as an offence against justice. But, it is surely not our sense of justice which is primarily outraged at the prospect of the demise of wildlife. The complaint against such a future is not that it would be an unfair one.

(b) Perhaps an analogous point could be made against the second position, but it is a different objection I will raise. Any diagnosis of our sentiment towards wildlife must take on board that most people would not

---

[3] According to an article in the *New Scientist*, the Director of London Zoo 'believes that the public is being misled by a barrage of animal welfare appeals disguised as conservation campaigns', whereas the Director of *Zoocheck* 'does not distinguish between animal welfare and conservation' (Macklin, 1990).

[4] Milan Kundera writes: 'Mankind's true moral test ... consists of its attitude towards those who are at its mercy: animals. And in this respect mankind has suffered a fundamental debacle' (Kundera, 1984, p. 289).

[5] Those, like Peter Singer, who do not like this particular vocabulary can, of course, equally deplore these evils, but in different terms.

equally regret the passing of the last tiger and the death of a member of a more numerous species.[6] Charles Addams' cartoon of two unicorns ruefully gazing at the disappearing ark would not work with two hyenas instead. So concern for a species is not equivalent to one for members *qua* individuals. It is not easy to grasp the exact nature of this concern, but one thing is clear: despite today's ubiquitous conservationspeak, few people are happy with keeping species permanently 'on ice', in zoos or safari parks. Even those *parvenus* masters of conservationspeak, zoo directors, must hold out to their public the prospect of one day returning their endangered charges to the wild. What matters to people, then, is the survival of a species *in situ*, in something like its natural state, not that of some confined, zoo-friendly *ersatz* species. But, there is nothing in the bare idea of a species' right to exist which can explain this aspect of conservationism: which means that it cannot, as it stands, capture the sentiment we are enquiring after. In practice, the imperative of species conservation tends to collapse into one or another version of the fourth position: species *in situ* are integral to ecological balance, perhaps, or promote the ideal of 'maximum diversity in unity'.

(c) Simple, honest introspection confirms, for many of us, that our regret at the disappearance of wildlife is not a mere function of concern for human well-being alone. Indeed, spokesmen for wildlife often testify to feeling sullied and compromised by the pressure, in our utilitarian climate, to focus on human benefits. If concern for the future of wildlife were really concern for our own, we would be much more depressed by a threat to such useful creatures as mussels and spiders than by one to creatures as dispensable for our comforts as dolphins and elephants. Someone will say, 'What about the aesthetic value to us of dolphins?'. But the aesthetic delight I have got from dolphins has been through watching films: so, if aesthetic pleasure is all that matters, then why, since there are plenty of these films in stock, should I worry if there are no more dolphins in the sea? After all, I do not worry over much that Fred Astaire is no longer with us, since what matters to me are his movies which, happily, I can still watch. Someone might, I suppose, say: 'Look, the sheer knowledge that there are dolphins in the sea gives you pleasure: that's why it's important to you that they be protected'. But this is a bizarre inversion of the truth. It is because it is important to me that there are dolphins, that knowing they are there gives me pleasure.

---

[6] Those, like Tom Regan, who dismiss all but the individual animal's rights, are therefore insensitive to most people's feelings about species. (See Regan, 1984, Ch. 9, Section 3.)

(d) The rhetoric of species conservation tends, these days, to merge smoothly with an environmentalist rhetoric. But the relationship between 'animal ethics' and 'environmental ethics' is a murky one. It is easy, certainly, to exaggerate their separation: after all, animals do or should have environments – portions of the world which are significant to them, and in which they can pursue meaningful activities.[7] (So I am puzzled when I hear of a forthcoming book on agriculture and environmental ethics which is not going to discuss at all the plight of battery hens or dry-stall sows.) On the other hand, unless one builds into the very definition of the 'health' of the environment that this environment should abound with, say, tigers and dolphins, it will be, at best, a contentious claim that their continued existence is ecologically essential. If it turns out not to be then, with the 'health' of the whole as the sole consideration, we ought not to feel any regret at the passing of these creatures. Stated differently: if the sentiment towards such animals were purely a function of a sentiment towards Nature-as-a-whole, it would. not be in the least offended by their demise. But, of course, it would. (Advocates of an exclusively 'holistic' ethic, like Aldo Leopold, tend to be very tough indeed on the fate of creatures deemed unhelpful towards general ecological welfare – to the point, indeed, that a critic like Tom Regan refers to their 'eco-fascism' (Regan, 1984, pp. 361–2).) To be sure, one might include among the criteria for environmental 'health' that the creatures we tend to care about most should flourish. But, in that case, environmental concern no longer explains, but is partly explained by, the sentiment towards wildlife manifested in that care. So either way – whether the survival of various species is or is not built into the notion of environmental 'health' – our sentiment towards wildlife cannot be construed as a function of a global attitude towards Nature, the ecosphere, or whatever.

None of the most familiar reasons given in defence of wildlife succeeds, then, in tapping the sentiment which inspires that defence. Before I work towards characterizing this sentiment, I need to correct an impression that talk of a human sentiment may have encouraged: namely, that it is a permanent, perhaps innate, feature of our psychology. We are dealing, in fact, with something that is rather recent. We care about the fate of wildlife in a way that our ancestors generally did not.

Here is one attempt to account for, and identify, the sentiment which has grown up. It might be called, after the title of McKibben's popular book (1990), 'the end of nature' hypothesis. The world is increasingly becoming

---

[7] See Cooper, 1992.

our world, a giant artefact. One-tenth of its land surface – and much more of its habitable surface – is cultivated or urbanized, and a good deal of the remainder bears a human footprint. Even the atmosphere bears the mark, notoriously, of our motor-cars and refrigerators. We will soon witness 'the end of nature', writes McKibben, where 'nature' is understood as what is 'wholly other', the counterpoint to human culture. And this he holds, is something terrible, for human beings need the presence of this 'other' from themselves and their products. Our burgeoning sentiment towards wildlife and the environment is the expression of this need. One may, no doubt, speculate variously on the source of this need. Maybe people need to experience a wonder or astonishment which their very own products, however impressive, cannot supply. Or, perhaps it is vital to them – and here the history of religion might be invoked – to feel dependent on, answerable to, something beyond the domain of their own activities and artefacts. Whatever the explanation, it is this need for the 'other' which underlies the concern for the wild.

But there, in those last two words, is the problem, for our purposes, with this account: it may explain a general concern for the wild, the whole natural environment, but not the more particular sentiment towards animal wildlife. That particular concern cannot, we saw, be subsumed under the general heading of environmental concern. Still, by calling attention to the galloping artificiality of the world and the unease which this occasions, 'the end of nature' approach starts in the right place. But we need to identify the dimensions of this artificiality which prompt our sentiment towards animal wildlife in particular.

Earlier in the century, people who predicted that technology would breed an alienated, rootless individual painted a nightmare scenario – Fritz Lang's *Metropolis*, Chaplins' *Modern Times* – of faceless clones chained to the production line. What has, in fact, materialized, however, is something gentler, but something to which people have yet to accommodate and something which is equally productive of rootlessness. I mean the erosion of the places which human beings once had in their environments.

Note the plural, for I am speaking not of the environment in which everything is located, but of the milieus in which people are (or were) 'at home'. An environment, in this sense, is not defined in geometric terms, as the area of x square yards or miles which surrounds a person. Rather, it is a field of significance for people: the domain in which they move confidently and with unreflective ease, in which they 'know their way about', and in which things and creatures have their place and meaning for them. Depending on the person's way of life, this field may coincide with a smaller or larger geographical space – from the smallholding of the traditional peasant to the oceans and ports of the 'China hand'.

It is possession of an environment which is eroded in modernity. The processes are familiar enough. Mobility, for instance, encompasses one type of modern process: a person is born in one place, goes to school in a second, college in a third, work in a fourth, fifth, and sixth, and retires in a seventh. And, even if people stay put, the traditional ties – a craft, say, or a chapel – which bound them together into a community has either disappeared or lost the strength to provide a sense of what matters and of how to behave. As the old is replaced by the new, moreover, the differences between environments are levelled out – everywhere the same supermarkets, restaurant chains, TV channels and housing estates. A person can now go far away and not be lost, and half 'at home' everywhere, but at the expense of no longer being fully 'at home' anywhere.

Technological modernity also transforms people's cognitive relation to their environments. The more technologized are the surroundings, the less relevant are the practical know-how, traditional skills, wisdom even, that people learned as apprentices from their elders. On the contrary, the knowledge that there is now a premium on acquiring may be a closed book to one's elders: the knowledge gleaned from manuals or courses at a college; the kind stored up and on tap in computers; and the kind needed to operate the gadgets that have rendered obsolete an older knowledge that was 'in the hands'. Then there is the impact of an education suited to a changing, mobile world, one which encourages a reflective, critical, even ironic distance from whatever environments have been inherited – environments which are no longer to be fitted into, or taken as given, but to audit, assess, improve, update, flit in and out of. This distance is then widened by a media industry which permanently displays to us a veritable *smorgasbord* of interests, lifestyles, values and tastes among which to pick, choose, and make up one's individual cocktail. A person has then become 'post-modern' man or woman. He/she is eclectic, pluralist, a citizen of the global village, a person, ultimately, who no longer inhabits an environment or world, but only the environment, the world.

Modernity and post-modernity are not, of course, without their admirers, and it is not my aim to debate the benefits which they may have conferred. But, it is important to recognize how the developments sketched above confront a very ancient ideal, one which, for instance, received beautiful expression in the classic texts of Daoism. This is the ideal of a smooth, natural, unreflecting immersion in a way of life where things, creatures, and places have their ready and familiar significance – a life in an environment where one is fully 'at home'.

There are many symptoms of the contemporary persistence of this ideal. For example, nostalgia for crafts, or admiration for the seemingly naive and innocent forms of life led by the few 'primitive' peoples that still survive. And here, too, I believe, is where to locate our current sentiment towards wildlife.

Two words which often figure in expressions of the sentiment towards wildlife are 'admiration' and 'harmony'. We are to admire wild animals, and especially for their seeming harmony with their surroundings. I am proposing, if you wish, a diagnosis or perspicuous description of this admiration: for what we admire (envy, perhaps) is, in my terminology, the wild animal's complete possession of its environment. Its harmony with its surroundings is no mere adaption to them – through camouflage, say, or ingenious thermostatic mechanisms. It consists, rather, in that unreflective ease of movement and activity, that practical knowledge 'in the hands' (or paws and claws) which, in a different modulation, is also what we admire in the traditional craftsman. As a living symbol of possession of an environment, the animal represents the antithesis of the technological artificiality of our present condition. More often than not, it belongs in a small and simple community where each creature has a natural place and role; absorbed in its environment, it is incapable of that ironic distance which we are increasingly incapable of closing; it fits itself into the rhythms of its life and the seasons, instead of endeavouring as we do to level them out; and its world is one, unlike our own, where things and other creatures have a consolingly stable significance and value.

Like any exercise in the phenomenology of an attitude, a suggestion like mine is hard to validate, and critics will charge that the diagnosis I offer may only describe my own, idiosyncratic sentiment. The final test, perhaps, is simply for the reader to try applying it to their own sentiments. But my suggestion has, I think some explanatory power: it lets otherwise puzzling phenomena fall into place. It can account, first, for the comparative recency, on a large scale, of the sentiment towards wildlife. Until recently, after all, people did not need symbols of a certain relation to an environment: they stood in it themselves. It can explain, too, the peculiar and proper hostility that people sharing the sentiment have towards the 'humanization' of wild animals – as performing seals, dancing elephants, tea-swilling chimps in bowler hats, and so on. Such stunts deny the animals' dignity, people say: and so they do – and this, I suggest, is because such animals are dispossessed of their environments and are enlisted into just those artificial practices which, in the wild, they symbolize the absence of and which, in human manifestations, is a source of unease at our contemporary condition. Finally, my suggestion can explain why some animals are more apt than others to evoke our sentiment. It is because some – dolphins, say, or tigers – are so obviously creatures that possess an environment, so visibly 'at home' with the surroundings through which they move with absolute ease. All animals, of course, must be pretty well integrated with their surroundings to survive at all; but, as symbols of this integration, some animals are better than others.

Let me avert a couple of possible misunderstandings. Although I emphasize the symbolic potency of animals in the wild, I do not at all count myself among those who condemn all keeping and training of animals by human beings. Today's dogs and horses are heirs of long traditions of domestication and life among humans, and in the better of those traditions, these animals, too, can possess environments. I agree with Stephen Clark that:

We humans can learn to see and live in beauty only if we acknowledge the real presence with us in the world and in our homes and workplace of creatures that are contributing members of our community.   (Clark, 1987, p. 176)

Secondly, my suggestion may seem to fall into the class of human-centred justifications for the protection of wildlife discussed earlier. But, it is not at all analogous to the argument that we should preserve wildlife because, say, it affords us aesthetic pleasure. Animals, indeed, afford us symbols, reminders, of something worthwhile – of a life congruent with the ancient ideal of integration with an environment. But we should protect them not because they serve as these welcome symbols, but because the lives they symbolize – lead, indeed – are to be admired. We protect them not in order to cater for a certain sentiment; rather that sentiment, properly diagnosed, intimates why they are worth protecting.

I have written at a level too general to yield specific proposals on educational and management policy. But in both areas, my discussion indicates broad directions.

There is a large consensus among environmental educators that concern for wildlife should be encouraged among young people, but the favoured strategies tend towards two extremes. First, an emphasis on the potential utility (medical, etc.) to humans of flora and fauna and, second, concentration on the contribution of wildlife to ecological stability.

The former is illustrated by remarks like these:

> by stressing the legitimate right of animals to live and survive free of fear and suffering, and thereby understating the value of such creatures to humans, animal rights advocates sometimes fail to raise some of the most compelling arguments in favour of wildlife preservation, ignoring points that may appeal to many otherwise unconcerned people.   *(Regenstein, 1985, pp. 131–2)*

This betrays a certain arrogance, for it is assumed that while 'we' committed aficionados discern the true value of wildlife, 'they' can only be impressed by appealing to their crassly pragmatic interests. Apart from the fact, already remarked upon, that many such appeals sound implausible, there now exists, if I am right, a widespread sentiment towards wildlife for education to both tap and help articulate. Teachers should not, therefore, be shy to encourage the young to recognize the value of wild animals, not in their practical 'pay-off', but as representing that beauty of an integration with a world which human beings have increasingly forfeited. And, outside of the school, 'we' should not be cajoled by hard-nosed media interviewers into supposing that 'they' can only understand the language of utility.

The second strategy – manifest, for example, in UK National Curriculum documents on environmental education[8] – is indicative of the excessive prestige enjoyed by a 'scientific' approach in educational circles. No one denies the importance of studying wildlife as parts of ecosystems, but this should not exclude considering other perspectives – pre-eminently, the animals' own. Teachers should facilitate understanding of animal behaviour in terms of how things are for the animals, of their goals, of what matters to them. For such understanding, we turn not to the scientist in his laboratory but to people who know animals in the manner of the forest-dweller, the observant hiker, or even the suburban family which attends to the wild visitors in its own back garden.

While the kind of education my discussion implies does not focus on utility and ecology, it is, of course, far from ignoring the value animals have for humans and the relations they stand in to environment. Indeed, it is precisely the animals' relation to their environments which is the source of the value we discern in their lives, since it recalls for us a dimension of an ideal human life.

Finally, a few remarks on wildlife management. Here, too, the favoured strategies tend towards two extremes. There is, first, the idea of 'the global

[8] See, for example, The National Curriculum Council, *Curriculum Guidance No. 7*, 'Environmental Education' 1990.

zoo': wildlife, if it is to survive at all, must do so within the precincts of zoos, parks, or 'worlds of adventure'. At the other extreme, we find the demand to preserve or recreate primal wildernesses, areas where animals may live with only minimal 'contamination' by human being, whether farmer, hunter, or tourist. The objection to the first strategy is obvious: the lives of wild animals matter to us precisely because they are wild and natural. The tiger in a cage – even one big enough for a land-rover to drive through – cannot be the symbol of a natural integration with an environment that its uncaged brother is.

Two remarks on the other extreme strategy are called for. First, given the pressures of poverty, population growth and tourism, it is utopian or frivolous even to insist that vast tracts of land in Africa or India – or England, for that matter – be set aside for the exclusive habitats of animals, out of bounds to all but a few privileged human visitors. The only complete wildernesses we are, realistically, likely to create are those where nothing lives – neither man nor beast: like those lunar landscapes left by giant herds of cattle which briefly grazed where tropical forests once grew. The most we can hope to achieve are relative wildernesses where human beings and animals coexist, perhaps in forest regions more gently and sustainably exploited than presently occurs in Sarawak or Brazil. No one, of course, pretends that the trick of promoting areas where humans and wildlife may exist in 'symbiosis' is easy to perform. And, some of the suggestions for how it might be done sound depressing: like a former Director of London Zoo's scenario of African villagers profiting from 'safari' hunts and the selling of animal souvenirs to the tourist hordes that pile through their lands.[9] But the trick must be brought off if wildlife is to be neither extinguished nor corralled inside a global zoo.

The second remark is that the creation of wildernesses, even if it were more practicable, is not obviously an ideal. For, if wildlife is to evoke the sentiment I described, it must be visible: some people must experience how wild animals live if they – and we, however vicariously – are to appreciate the lives of these creatures. Thus – to take some small-scale but telling illustrations – one should welcome, and not dismiss or only grudgingly concede such efforts as *Into the Blue*'s support for areas where people may 'swim with dolphins' that themselves choose to 'swim with humans'; the *Fox Project*'s encouragement of a controlled presence of foxes adjacent to our everyday lives; or the Ranthambore Society's revival of traditional Indian crafts and agriculture that do not intrude, beyond necessity, on the

[9] 'The Society and Conservation in Africa', Memorandum, 9/1/91.

wildlife in the surrounding landscape. These are undramatic illustrations of human beings incorporating into their lives an experience of animals that adds to those lives. They do not illustrate something 'second best', an unfortunate but necessary accommodation with a harsh, modern reality: rather, they recall a relationship to animals that was once a natural feature of human existence. Indeed, it is only with the erosion of that kind of relation, and the experience of this as a loss, that the future of wildlife can be appreciated by us as an 'issue' or 'problem' at all.

Acknowledgement

We are grateful to the editor of *Environmental Values* for permission to reproduce a revised version of this chapter.

References

Clark, S. R. L. (1987). Beasts like us. *Times Literary Supplement*, 20/2/1987.
Cooper, D. E. (1992). The idea of environment. In *The Environment in Question*, ed. D. E. Cooper & J. A. Palmer, pp. 165–80. London: Routledge.
Kundera, M. (1984). *The Unbearable Lightness of Being*. London: Faber & Faber.
Macklin, D. (1990). Who's hijacking conservation?, *New Scientist*, 8/9/1990.
McKibben, B. (1990). *The End of Nature*. Harmondsworth: Penguin.
Regan, T. (1984). *The Case for Animal Rights*. London: Routledge.
Regenstein, L. (1985). Animal rights, endangered species and human survival. In *In Defence of Animals*, ed. P. Singer. Oxford: Blackwell.
Rolston III, H. (1986). *Philosophy Gone Wild*. Buffalo: Prometheus.

17

In the absence of animals: power and impotence in our dealings with endangered animals

CHARLES BERGMAN

I was not supposed to see what I saw. It had taken me a month of persistence and all the subtleties of deference to get access to these condors, to be in this place at this time. It made me a witness, not to a nicely managed public-relations triumph, but to a secret no one wanted to admit.

AC-3 was dying. We didn't want to admit it. But no amount of wishful thinking could alter what would turn out to be her end. She was the last California condor to die in the wild. She did not precisely die in the wild, but she might as well have. She was the last successfully breeding female condor in the wild, and by the time she was captured, she was nearly dead anyway.

I saw her in the wild just before she was captured, before anyone knew for certain what was wrong with her. I watched her for a full day. She sat across the canyon from me in the top of a pine, behaving strangely.

Her perch was about half a mile from us, across a steep ravine called Bittercreek Gorge, pleated with ridges and alive with the splendid greens of pines and live oaks. On the top of a pine, AC-3 would shake out her wings, immense and black with white triangles in the lining – the classic identification mark for a condor. A huge vulture hunched on a melancholy perch, she betrayed no hints as to what was wrong with her. The slanting winter sun glanced off her black body with dazzling iridescence, the liquefaction of a solid black bird. Her head was featherless, and in the honest brightness of the morning, it looked pink, shading to dull oranges and scaly flesh tones. She looked like a person who had held her breath too long, and grown red-faced and puffy. With her huge beak closed, she seemed to have a sort of pursy, senile smirk on her face.

For over a week, AC-3 – short for Adult Condor 3 – had not left her roost. Her long-time mate, AC-2 sat below her in the same pine. They were the last successfully breeding pair of California condors in the wild, and for several years, as condors kept plunging toward extinction, this pair had been among the big producers, a reliable old couple. After half a century of environmental rescue efforts, and after 5 years of intense biological triage for the species, we were reduced to this last breeding pair, and four additional wild condors. Because of their rarity, these condors had grown increasingly precious, as life always does when tragedy grows more imminent. And now, AC-3 was not flying. She had not left Bittercreek Gorge in over a week. Condors always flew. In the past, this pair would have flown 100 miles easily from roosting to foraging places.

Only three days later, the biologists decided to catch AC-3. As with condors in general, they had given up hope that AC-3 could recover in the wild. They captured her on January 3, 1986, and took her to the San Diego Wildlife Park. She was so weak that the ornithologist who captured her just walked up and grabbed her. Tests confirmed the worst fears: high levels of lead in her system, present for at least 2 months. Lead in the system – usually picked up from pesticides or bullets in carcasses – freezes up the peristalsis. She had been eating in the past weeks. Some of the food in her engorged crop was 2 weeks old. Since she could not digest, the food sat in her mouth and rotted.

AC-3 was scrawny, starved, and dying. At the zoo, veterinarians laid her on a white antiseptic table, stuck plastic tubes in her for intravenous feeding, and pumped her system repeatedly. Biologists and veterinarians tried desperately and futilely to clean her out. Fifteen days after her capture, on January 18, 1986, AC-3 died.

That one day, sitting across from an immobile AC-3, altered the way I viewed the phenomenon of endangered species in North America. Through the languors of a warm winter afternoon, I had the eerie sense that I was the vulture, watching the condor dying, and the real vulture was the victim. We were impotent in our attempt to save her. With endangered species more generally, we have become witnesses to the story of death and loss happening right before our eyes, as we gaze on, helpless. The gaunt animals are gathering outside our houses, and the corpses are piling up. I have come to believe our strategies for saving endangered species are in any larger sense doomed to failure, despite all our good intentions. Why? The limits of our attempts to save endangered animals can be summarized succinctly in our concept of wildlife management – which is, after all, a contradiction in terms. It will necessarily lose what it sets out to save.

In their imperilled status, endangered species represent a paradox: although they are the result of our long obsession with power over nature, they embody the limits of that power. They are images, not of our stunning triumphs, but of our failures. I no longer view the phenomenon of endangered species as simply an unfortunate accident. They are an inevitable consequence of our relationship to animals. We have tried to define the problem of endangerment as primarily a bio-political problem. We see the animals largely as creatures to be manipulated in one way or another, through maintaining minimum viable populations, habitat preservation (largely in National Parks and National Wildlife Refuges), captive breeding programmes, and animals in zoos (Ehrlich & Ehrlich, 1981; Myers, 1979; Lovejoy, 1986; Vermeij, 1986). These techniques for saving endangered animals cannot succeed, in any larger sense, because they derive from the same mentality, the same posture toward nature, that created the problems in the first place.

Furthermore, these broken creatures have become less a part of nature than of our culture. We made them, we preserve them. We like animals because they are endangered. Increasingly, being endangered confers status and prestige upon an animal – it constitutes in itself the significance of a animal for us. There is something troubling in this for me, as if somehow we have appropriated the phenomenon of endangered species, incorporated it into our lives. Despite what we say about our feelings for endangered animals, I fear that through our reverence for them we are learning to live in the absence of animals. Extinction, like the poor, may always have been with us, but endangered species are a modern invention, a uniquely modern contribution to science and culture.

Whether rendered explicit, of left implicit, our approach to the problem of extinction begins with the question, 'Why save endangered species?'. The question makes me mad, partly no doubt because there is no satisfactory answer. But it is also the wrong question.

The best answers derive from the theoretical essays of Aldo Leopold. The environmental movement, and the effort to save endangered species, is founded largely on his work. Though the movement is older than Leopold, he gave it both ground and impetus in 1948 with his justly famous essay, 'The Land Ethic' (Leopold, 1949). In it, he argued that we need to see the land – which for him was the interlocking web of animals and plants and soil – as a community of which humans are a part. 'In short,' he wrote, 'a land ethic changes the role of *Homo sapiens* from conqueror of the land community to plain member and citizen of it (p. 204).' We have come to think of this concept as 'ecology'. Aldo Leopold was part of the great effort to break down the ancient western dualism between humans and nature. He helped to show us our place amongst the beasts.

On the theoretical level, this concept is limited by its literalism. The land ethic, and the environmental movement built upon it, still identifies creatures as something out there, as external to us, separate and apart. But stewardship is itself an arrogant stance toward nature, a patronage that betrays our limited view of animals even in our concern for them. They remain for us something to control. From our exalted position, we will take pity on the less fortunate creatures, the ones unable to adapt to life on the terms we have set: the fast-paced change and sweeping destruction of the earth, both of which nullify hopes of evolutionary adaptation. Stewardship is noble, no doubt, and better than dismissing the creatures. But our concern does to endangered animals what it also has done to the poor. It turns them into failures. They aren't as good as we, not as strong nor as successful, but, we think smugly, we'll give them a hand anyway.

The worst reasons for saving endangered animals are probably the most effective: the ones that are blatantly, distressingly utilitarian. Although they galvanize the most support for saving species, such reasons are just endangered species, so the argument goes, because we never know where the cure for cancer will come from.

In the politics of preservation and extinction, these arguments have not really worked, either. The United States has undertaken the most progressive and ambitious attempt of any country to save endangered species. Our concern has been codified into national legislation and bureaucratic procedures. Our protection has, of course, resulted in some notable successes. The American bald eagle – the national symbol – is one example of a species saved by our concern. Though it suffered a slow decline in numbers as Europeans spread westward across the continent, it went down fast in the 1950s because of DDT. Three decades of protection and active management have produced a comeback. Its current population in the lower 48 states is estimated at between 7500 to 10 000 animals. The United States Fish and Wildlife Service has reclassified the American bald eagle from endangered to threatened. The population, however, is not anywhere near estimates of its original numbers, which have been estimated at anywhere from 25 000 to 75 000 birds.

Yet, before we rush to self-congratulation, we should realize that this success is largely symbolic. Listen with your ear cocked, and you can hear the cultural self-vindication in the rhetoric of the recovery – not just for the nation, but for the wildlife industry as well: 'This is a good sign for endangered species recovery,' as one person said. 'There has been so much money and effort, and this bird is the national symbol.' Maybe this is the language by which we are trying to salvage some hope. Even biologists working on behalf of individual species realize that we have changed the landscape so profoundly that the future holds little room for wildlife.

So while we have saved remnants of some of the spectacular species of the last few decades, the problems facing wildlife continue to escalate. As of October, 1995, there were 752 US species or subspecies endangered on the federal list and 206 threatened on the list. Making the list is the crucial achievement for any species in trouble. Still, the backlog of species which are candidates for listing, but which have not been approved, has remained constant for about 15 years: between 3000 and 4000 species. Nothing can be done for these species, except public agitation, until they are listed.

But listing is no guarantee of recovery. A Government Accounting Office report in 1989 estimated that of all the US species, only 16 could be considered recovering. Another 18 species are now thought to be extinct, though they very well may have gone extinct before the passage of Endangered Species Act. According to many biologists, we are quite literally living in an age of mass extinction, far beyond anything we associate with that other staggering age of extinctions, the Pleistocene (Lovejoy, 1986, p. 16; Myers, 1979, p. 4).

At best, we have achieved important but limited successes, refuges of survivors in a storm that continues to howl all around. Somehow, the question, 'Why save endangered species?' presumes that it might be all right to lose the animals, or that endangered animals might even be a given, something we have to learn to live with, one of the 'data' of our lives. The question presupposes that we are the lords of creation, and that it is our right and duty to oversee nature. It derives from the same sense of superiority to the nature that created the problem in the first place. In this context, work on behalf of animals is always a sad rearguard action, and we are reduced to such feeble but well-meant gestures as putting condors into zoos to save them. Meanwhile, the wreckage worsens.

I am genuinely happy about the recovery of the bald eagle. But I am also sad about the circumstances facing smaller, more nondescript creatures. The most compelling juxtaposition of our power with our impotence lies for me in the dusky seaside sparrow. In 1979, biologists undertook a project to save dusky seaside sparrows – it was beautiful and quixotic and doomed from the start. They captured all of the final dusky seaside sparrows they could find – five males. Only males. They tried for a decade to crossbreed them with a near relative, the Scott's seaside sparrow, so that when the males died, the genes of the species would live on. I saw the last of these males, named Orange for the colour of his leg band, in his cage. I was aware when I entered his cage that I was practically looking at extinction. I was as close as you could come to it. Orange was the last of his species – solitary, senescent male, living out his final day behind wire mesh. He died on June 16, 1987. He was found in the morning, keeled over in his water dish. His cage? It's well known, but I still have trouble wrapping my mind around this: the last dusky seaside sparrow died in that monument to American fantasy and wealth, Disney World, in Orlando, Florida. Extinction in the midst of the Magic Kingdom.

I think the question we have asked – 'Why save endangered species?' – is the wrong one. The question I want to ask about endangered species reverses the customary point of view. I want to ask, 'What do endangered species mean?' Why do we, in our time, have this epidemic of endangered creatures? What do the lives of these broken creatures say about us and our relationship to nature?

Of course, they indict us. But the indictment cannot be for a lack of concern and effort on their behalf. It is not that we have not tried to save endangered species. I can best explain what endangered species expose in us by using the example of the California condor.

In many ways, the failing condition of AC-3 was a summary of her entire species. For over a century, the condor had followed an ineluctable decline. On Tuesday, December 17, 1985 – the day that I was watching the sick AC-3 in the field – the US Fish and Wildlife Service announced that it would capture the last six California condors in the wild. Even as I watched AC-3, wondering what was wrong with her, I knew I was seeing the end of these birds in the wild. The decision was not a giving up. It seemed then, and it still does, the last best hope for the species. Yet, despite the optimism that was attempted at press conferences, having to capture the condor was anything but a triumph. It was impossible not to feel that something had happened that we could not control, that some impotence in us was exposed.

What was also exposed in the decision, however, was a new way of responding to endangered wildlife. In the defeat of the wild condor, a new condor has emerged. With it, we have crossed over into a new posture toward wild animals, contained within a highly manipulative biology. The strategy of capturing extremely endangered animals has led to a new category of creature: species that are extinct in the wild, but which we will try to save in zoos. The dusky seaside sparrow died under such circumstances. The last Guam rail was captured on March 1, 1985, and the last black-footed ferret was captured in February 28, 1987. The strategy with respect to the California condor is revealing. At the time the decision to capture the last six condors was made, California condors had never been successfully bred in captivity. That was part of why the decision to 'bring them in' was so desperate.

The most startling and controversial part of the plan to rebuild the species from a captive flock, however, is the programme to release the condors after the captive breeding has produced a genetically varied zoo flock. On the basis of work he has done with Andean condors in Peru and California, the biologist who developed this plan, Michael Wallace, with the Los Angeles Zoo, was convinced he can train condors to live safely in the remote wilderness (Wallace, 1983). Already, female Andean condors were being released in the mountains north of Los Angeles in surrogate species experiments as the final California condors were being captured. (The release of only one sex prohibits reproduction by the species in the area.) California condors may be released within a few years – sooner than formerly anticipated. The plan was to release California condors and control them through the placement of carcasses. Condors transmit strong 'traditions' to the young, and through 'carcass management', biologists will instil in the species new traditions to keep them in safe habitat. The introduced Andean condors fly widely, but they feed only on the carcasses placed out for them. As the biologist said about running the programme from the Andean condors, he has 'very good control over them'.

Even when the animals that were captive bred are released, they will be genetically like their wild parents or grandparents, but behaviourally, they are very very different. What is lost when animals are kept in captivity, and cannot be maintained or passed on, even when released, are their natural behaviours. This has been found distinctly from the release programme with the golden lion tamarins, which in most ways, is a large and expensive failure. Released tamarins do not tend to live long, however, if their offspring do survive, then it is only third generation releasees that show survivability. In fact, the initial released animals don't even know how to locomote or find food in the simplest way. They are utterly helpless at surviving in the wild, as we would be, if we were suddenly dumped in deep rain forest without tools, knowledge, and experience.

Yet, with all these complex doubts and reservations, and after all the paroxysms of anger and grief about capturing the California condors, the captive breeding and release programme has been much more successful than people had hoped. It remains a complex and expensive and qualified success, but there are now California condors back in the wild mountains north of Los Angeles. The first successful birth of a California condor in captivity came on April 29, 1988. Named Mollokom this baby signalled the first real hope for the species. By 1990, the number in captivity had reached 32. Then, on January 14, 1991, the first captive-bred condors were released back into the wild. The pair were Xewe, a 9-month old female hatched at the Los Angeles Zoo, and Chocuyens, an 8-month old male hatched at the San Diego Wild Animal Park. They flew out of a pen in the Los Padres National forest's Sespe Condor Sanctuary. They had lived in this holding pen or enclosure for three months on the side of cliff before release (Haas, 1991). As of October, 1995, there are 11 birds flying free in the Sespe Condor Sanctuary, near where I had seen the condors just before all of them were captured. Since then there have been five releases of condors. The total number of birds in the wild varies, almost on a weekly basis, since they are closely monitored by the US Fish and Wildlife Service (Wallace, 1991). Birds that need care or attention are captured and returned to the zoos, four birds have died after release, as well. As of October 1995, 13 birds are in the wild out of a total population of 103 California condors.

Condors in a zoo. Condors on a leash. Condors were once the symbol for us of an entire nation's commitment to, and concern for, endangered species. With this new strategy, we have been forced to acquiesce to a notion of creatures as objects to be manipulated. Although I have a streak of the romantic, I am not advocating a nostalgia for a lost innocence or purity here. I do want, honesty, however, about what we are doing. We are saving creatures largely through a strategy of confinement and control, and not just zoos. Even creatures in the wild are increasingly relegated to a more invisible form of confinement: national parks and wildlife refuges. This is another sign of the increasing alienation of our culture from the parts of nature that reason cannot control. Animals are allowed to live merely on handouts along the borders of our lives. In the condors, I find a metaphor for the marginalization of wildlife in the United States, even as we save it. We are according an official, marginal status to wildlife. The spatial sign of this is a zoo or park. The verbal sign of this is the category, 'endangered'.

It is hard not to conclude there is something convenient for us in this arrangement. More worrisome, there seems to be an unconscious collusion going on between environmentalists, scientists, and society. It may be a collusion of good intentions, but power often wears a mask so that it conceals its own operations from itself. We want to save the condors, we say, but we do not admit that this actually gives us complete control over the creatures.

With the decision to place all the condors in the zoos, one of the winners was biology. What it has won is the right to control the condor. The science of biology, increasingly, has won the right to tell us what a creature is, to define an animal for us. Yet our science of creatures conceals from us the way we exercise power through it. Everywhere in his work, the French philosopher Michel Foucault (*Sexuality*, 1978) exposed and attacked and studied this substrate of power under our knowledge. The 'rational' and 'empirical' are what Foucault calls the 'governing codes' of our culture, expressed through our science and used to give our particular form of order over the world. But there is an 'unspoken order' behind these codes, which they cannot articulate. In *The History of Sexuality* (1978) Foucault finds beneath our rationality the 'omnipresence of power'. It is, he says, 'not an institution, and not a structure; neither is it a certain strength we are endowed with; it is the name that one attributes to a complex strategical situation in a particular society' (p. 93).

We should be able to ask what deeper objectives are served by the kind of knowledge we generate about nature. What kinds of domination are served by our knowledge? In what ways has science become the ruse of reason? One of the founders of modern science, Rene Descartes is explicit about the goals of scientific knowledge. In 'Part Six' of the *Discourse on Method*, he argues that we should know the 'power and affects' of the natural world so that 'we might put them ... to all the uses for which they are appropriate, and thereby make ourselves, as it were, the masters and possessors of nature' (p. 45).

In the case of the California condor, biology is part of a larger strategy as well. It exists not as pure knowledge, but as part of a means to give us control over the bird. It is in this way that I find the case of the condor symbolic of what is happening more broadly with endangered species. Our response to endangered animals has been a classic example of American pragmatism: define a phenomenon as a problem that we can do something about, and then invent a solution. The focus on a single species, the emphasis on identifying problems, on generating solutions, and on extending out control over nature – these are the forms of power as expressed in our current approach. Yet, this very desire to exert power has informed our society since at least the 17th century. The methods we have used to help endangered species cannot work, except in isolated cases, because they are part of the same mentality and they serve the same ends that created endangered species in the first place.

At the same time, however, we are transforming creatures. In the last few centuries, we invented animals as biological creatures. In the last few decades, we have re-invented animals as biological problems. Subjected to the control of reason, they have been transformed into a strange mechanism. This is the central paradox: at the same time that we demystify the creature through our science, we lose touch with its secrets. Putting animals in a zoo can save them – a strange formula. We confirm in that act our alienation from the 'beast'. We ensure that it is no longer inside us, but living in a separate, designated space. The beast is isolated. Through our science, we give to the beast, not its expression, but its cure.

The condor shows the double face of our science. Where there is power, there is also impotence. Where there is light, there are also shadows. We have concealed parts of the beast from ourselves in the process of disclosure. Endangered animals are one of the consequences of our power over nature. They are images of the unconscious of science, the dark side of our relationship with nature.

Reluctantly, I came to support the plan to capture the condors and place all our hope for their future in the captive-breeding programme. It has proved successful, but highly expensive and useful only for the most high-profile species. Desperate measures are required for animals that are as severely endangered as the condor, or the black-footed ferret or the dusky seaside sparrow. This is especially true, because the lives of individual animals like AC-3 or Orange are always at stake. But we should not delude ourselves about what we are doing. Even as environmentalists and biologists fight for the animals, we may be co-opted in our guerrilla victories, swallowed by the larger forces at work. What, after all, is the liberal environmentalist's dream? He/she believes that five more miles to the gallon will save the planet. He/she believes in technology, because it gives him/her all those beautiful nature shows on endangered species on his/her Sony TV. He/she helps the bureaucrats divvy up the earth so we can slow down the growth in the rate of destruction. And without realizing it, he/she has come somehow to believe that the diminished beast is the real beast.

Though scientists have documented our environmental problems, it is naive to believe, as we have for the past 50 years, that they can solve them for us. They are implicated by an ideology of power over and distance from nature before they begin. What should we do? Perhaps, nothing. Perhaps, we are not yet ready to solve these problems. That is, what we do at this point may be less important than what we think and feel. The need for action, the pragmatic impulse, is so deeply ingrained in the western psyche, we have trouble realizing that the path into the future may not lie in what we do, as such, but in our ability to re-think the animals. And, in so far as Americans and Europeans develop the environmental models emulated by the rest of the globe, we need to be especially cognizant of the attitudinal and philosophical underpinnings that have caused the epidemic of extinctions that characterize our times.

In the context of an increasing distance from nature, both literally and emotionally, in a landscape increasingly defined by absence, we need a more intimate concept of ecology of humans and animals. At this point, we do not really know what might be, and perhaps we need to feel more fully the loss of animals before we are able to re-think them. Some day the earth may be so bare and depleted we will no longer be able to ignore the consequences of our lives. In the meantime, these lost animals are the outward images calling us to the internal struggle for new hearts and new minds. The task we face to re-create animals by learning to re-image them and our world.

The beginning of this process is to realize that an animal is only partly a biological creature. The science of biology tells us what an animal is. But, an animal is also a symbol. Seeing an animal as a symbol tells what it means. Seeing an animal as a symbol enables us to understand a creature's emotional significance in our lives. It shows us the place of the animal inside, the creature of our imaginations. It tells us about ourselves. How we relate to nature depends on how we see it, and part of the reconstruction of nature depends on the reconstruction of our own ways of seeing. The symbol enables a community to explore its relation to itself and its environment in ways not possible through the objectivity of science. I have one experience with condors that suggests the loss we have suffered as animals vanish. It also suggests how, through symbols, we can recognize a different relationship with creatures.

In those last days of the condors in the wild, I arranged to join the pilot who tracked the remaining condors through radiotelemetry from a small plane. We were going to follow two condors, AC-8 and AC-9, from their foraging area to their roost deep in the Sierra Madre Mountains. We picked up the birds on the Tejon Ranch, east of Interstate 5, and followed them from a discreet distance as they made their way west. They reached the freeway at the 'Grapevine', a long grade rising into the mountains just north of Los Angeles. The freeway is cut deep into the mountains. By late afternoon, the artificial canyon of the freeway surges with radiant heat, bouncing thermals, and slicing winds from the ocean. These currents must have been the stuff of dreams for condors. AC-8 and AC-9 did not cross the freeway. Instead, they began circling, wide and slow, ten-foot wings held steady, tilting into the turns. One of the condors rolled left, the other rolled right, spiralling upward in a double helix upward through the billowing air.

In our two-seater Citabria, we leaned into a tight spiral, too, and rose with the condors. The small plane insinuated itself into the condors' flight with an intimate aerobatics. Slowly, almost deliberately, like partners in a dance that is unrehearsed but unfolds with a kind of inevitability that looks calculated, all of us flew higher, leaving the freeway behind, forgetting the cars below pouring in and out of Los Angeles. That was the culture below, and it seemed suddenly earthbound, oblivious to the drama unfolding in the skies above. We stayed closest to AC-9, who was trailing AC-8. We kept circling with the condors, rising above the highest mountains, the views getting more expansive, the spirals growing wider in a dilating euphoria.

The California Indians revered the condor. They conferred upon it the power to transport humans into a different dimension. Many tribes celebrated a festival called *panes*, in which a condor was sacrificed to a god and its skin removed in one piece, was buried with seeds. The condor was thought to be transcendent, able to come back to life. In fact, the North Fork Mono Indians believed the same condor was sacrificed every year. In one tale, the condor was so powerful that he would carry people who slept out in the open up into the skies. He would set them free in 'Skyland'.

With these condors, I was in their element. I do not normally prefer to see wildlife from an airplane. It is the easiest way, usually, to see rare animals, but it is distant, impersonal. I was very aware of the incongruity of me in the plane, peering out the windows, watching AC-9 cruise on the air. But I was seeing the condor in a new way. He was a magnificent soarer. And, he lifted me, as in the Indian tale, into a higher space, a paradise of flight and wide-open skies.

What I did not know then was that AC-9 would be the last condor to be captured and brought into a zoo. It took the biologists over a year to catch him. He was finally caught on Easter Sunday, 1987, the last wild condor. No matter what happens with condors in the future, my memory of him is made poignant with the sense of loss. We have lost more than just numbers of animals. Our domination has despoiled them of their power – to lift us, to inform our imaginations, and even to teach us about our lives. Try as we might, however, we cannot separate ourselves from the animals. Even their extinctions come back upon us in a growing void. These lost and broken creatures are reminders, that the more complete our domination of nature has become, the more we have lost in the process.

At several thousand feet, the condors banked out of the spiral and headed south, wings firm, rarely flapping, in a 'flex-glide'. In the plane, we stayed about 100 feet off the wing tips of AC-9. His wingspan was almost a third of the size of the Citabria's. The day faded in a subtropical swoon of rich light – velvety reds, dusky peaches, heady yellows, a tinge of purple that almost hurt to behold. The late rays blinked off the condor's black body in a melanistic sheen. His head swivelled, the way a modern dancer will isolate and move a single body part. I stared out of the plane window, watching his red head rotate while he flew, looking below, looking sideways. Looking straight at me.

There is nothing like being transfixed in the gaze of a spectacular wild animal. It is always a shock for me – the sudden recognition of strangeness. In the AC-9's stare, I felt a disorientating self-consciousness that came with losing my role for a moment. I was no longer sure whether I was the seer or the seen. I got a fleeting sense of what I must look like to the condor, both of us made visible in the same light of day.

The two condors turned off the freeway after several miles and made towards their roost. They would slide low along a ridge, connected it seemed to the contours of the hills. On the sway of the winds, they had an unbelievable grace, indolent and unhurried. At Agua Blanca, they again began to circle, staying just above the shadows, reeling in the fiery light. Now and then, they would slip out of the light, down a ridge, lost in a black swatch, only to catch another thermal and burst back into the light, like buoys breaking the surface of dark waters.

AC-8 and AC-9 spun together in lovely pirouettes above their roost. The sun burned along the horizon, and the condors hovered just above the canyon. The late rays splashed off their broad wings. Emblems of their species, black bodies polished and brilliant and unforgettable, the two condors hung just above the darkness below.

References

Anon (1988). California condor population grows by one. In *Endangered Species Technical Bulletin*, **13**, 1.

Descartes, R. (1637, 1960). Discourse on Method. In *Discourse on Method and Meditations* (trans. Laurence J. Lafleur), pp. 3–57. Indianapolis: Bobbs-Merrill.

Ehrlich P. & Ehrlich, A. (1981). *Extinction: The Causes and Consequences of the Disappearance of Species*. New York: Random House.

Foucault, M. (1978). *The History of Sexuality, Volume 1: An Introduction* (trans. Robert Hurley). New York: Random House.

Haas, A. (1991). History on the wing: California condors restored to home skies. In *Endangered Species Technical Bulletin*, **16**, 1–15.

Leopold, A. (1949). The land ethic. In *A Sand County Almanac and Sketches Here and There*, pp. 201–26. New York: Oxford University Press.

Lovejoy, T. (1986). Species leave the ark. In *The Preservation of Species: The*

Value of Biological Diversity, ed. B. G. Norton, pp. 13–27. Princeton, NJ: Princeton University Press.

Myers, N. (1979). *The Sinking Ark: A New Look at the Problem of Disappearing Species*, New York: Pergamon.

Odgen, J. (1985). The California condor. In *Audubon Wildlife Report, 1985*, ed. Roger di Silvestro, pp. 388–99. New York: Audubon Society.

Vermeij, G. (1986). The biology of human-caused extinction. In *The Preservation of Species: The Value of Biological Diversity*, ed. B. G. Norton, pp. 28–49. Princeton, NJ: Princeton University Press.

Wallace, M. (1983). Evaluation of techniques of hand-reared vultures to the wild. In *Vulture Biology and Management*, ed. S. R. Wilbur & J. A. Jackson, pp. 400–23. Berkeley: University of California Press.

Wallace, M. (1991). Methods and strategies for releasing California condors to the wild. In *American Association of Zoological Parks and Aquariums, Annual Conference Proceedings*, pp. 121–8. San Diego, CA.

Index

1918364   Ni